Geology

250

MESQUITE

US/IBP SYNTHESIS SERIES

This volume is a contribution to the International Biological Program. The United States' effort was sponsored by the National Academy of Sciences through the National Committee for the IBP. The lead federal agency in providing support for IBP has been the National Science Foundation.

Views expressed in this volume do not necessarily represent those of the National Academy of Sciences or of the National Science Foundation.

US/IBP SYNTHESIS SERIES | 4

MESQUITE

Its Biology in Two Desert Scrub Ecosystems

Edited by

B. B. Simpson
Smithsonian Institution

Dowden, Hutchinson & Ross, Inc.

Stroudsburg   Pennsylvania

79 78 77   1 2 3 4 5
Manufactured in the United States of America.

Library of Congress Cataloging in Publication Data

Main entry under title:
Mesquite, its biology in two desert scrub ecosystems.
  (U.S./IBP synthesis series ; 4)
  Bibliography: p.
  Includes indexes.
  1. Mesquite. 2. Desert ecology. 3. Botany—Arizona—Tucson region. 4. Botany—
Argentine Republic—Andalgalá region. I. Simpson, B. B. II. Series.
QK495.L52M47   583'.322   76-58889
ISBN 0-87933-278-6

Exclusive distributor: **Halsted Press**
A Division of John Wiley & Sons, Inc.
ISBN: 0-470-99109-7

to the memory of

ARTURO BURKART
1906–1975

FOREWORD

This book is one of a series of volumes reporting results of research by U.S. scientists participating in the International Biological Program (IBP). As one of the fifty-eight nations taking part in the IBP during the period of July 1967 to June 1974, the United States organized a number of large, multidisciplinary studies pertinent to the central IBP theme of "the biological basis of productivity and human welfare."

These multidisciplinary studies (Integrated Research Programs) directed toward an understanding of the structure and function of major ecological or human systems have been a distinctive feature of the U.S. participation in the IBP. Many of the detailed investigations that represent individual contributions to the overall objectives of each Integrated Research Program have been published in the journal literature. The main purpose of this series of books is to accomplish a synthesis of the many contributions for each principal program and thus answer the larger questions pertinent to the structure and function of the major systems that have been studied.

Publications Committee: U.S./IBP
Gabriel Lasker
Robert B. Platt
Frederick E. Smith
W. Frank Blair, Chairman

PREFACE

Few plant genera of the New World deserts have received as much attention as *Prosopis*. This special interest is due, in part, to the fact that several species of *Prosopis* are serious rangeland pests in both North and South America. However, in addition to their importance in modern man's economy, trees of *Prosopis* play a major role in the functioning of New World warm desert ecosystems. From 1971 to 1975, the Origin and Structure of Ecosystems Project of the International Biological Program carried out research on convergent evolution in warm desert scrub ecosystems. Two disjunct American sites were chosen for intensive study: one in the Sonoran Desert near Tucson, Arizona, U.S.A., and the other in the Monte Desert near Andalgalá, Catamarca, Argentina. During the period of this study, it was apparent that species of two genera dominated both study areas. One, *Larrea*, is the genus of creosotebushes, overwhelmingly dominant in New World deserts in terms of numbers of individuals and geographical area covered. The second, *Prosopis*, contains the mesquites and algarrobos. Although individuals of *Prosopis* are much less abundant than those of *Larrea*, mesquite can be considered dominant in terms of its influence on the biota of these warm desert ecosystems. Throughout the study of convergence, therefore, much of the work perforce centered around species of these two genera.

In this volume, contributions of twenty-two scientists come together to produce a picture not only of the role of mesquite in the lives of the modern biotas as well as past and present human cultures around the study areas, but also of the physiological, morphological, and genetic mechanisms present in species of *Prosopis* that allow these taxa to survive under arid and semi-arid conditions. A comparable synthesis with *Larrea* as the focal point has also been prepared (Mabry et al., 1977).

For each chapter, a coordinator, the first author listed, collated the contributions of the chapter authors, who are listed alphabetically. Within each chapter, when applicable, footnotes designate the authors of particular sections. For these sections, the order of the authorship was agreed upon by the authors themselves.

In addition to the authors, however, this synthesis reflects the help and support of many other individuals and of several institutions. In Argentina, the Instituto Miguel Lillo in Tucuman provided indispensable facilities. Drs. Marta Grassi, Peter Seeligman, Lionel Stange, Federico Vervoorst, and Abram

Willink gave generous and continuous help and advice. In Arizona, the University of Arizona Department of Biological Sciences provided laboratory space and equipment for several years. Drs. Charles Lowe, Charles Mason, and Tien Wei Yang provided assistance and encouragement.

Funding for the individual research projects was primarily provided by the National Science Foundation through the International Biological Program and included grants to Drs. Frank Blair, Tom Mabry, Andrew Moldenke, Harold Mooney, Gordon Orians, and Otto Solbrig as Principal Investigators. Support for the work summarized in Chapter 8 was provided independently by the National Science Foundation under grant SOC 75-13-628 to Richard Felger. Most of the research of the editor was funded by the Smithsonian Institution, Department of Botany.

The critical reviews of the manuscript by Drs. James Brown, Herbert Hull, Paul Martin, and Talmar Peacock led to substantial improvement in the text for which the authors and editor are grateful.

The impetus the Origin and Structure of Ecosystems program provided the scientists involved, many of whom were students at the time of these studies, has led to continuing research in a wide spectrum of fields. We have all benefited from the exchange of ideas that took place during this collaborative effort.

B. B. Simpson
Smithsonian Institution

ACKNOWLEDGMENTS

Many persons have directly or indirectly helped in the preparation of this volume. The editor would like to thank M. J. Johnston for help throughout and especially E. L. Davis for typing and S. Yankowski for help in preparing the manuscript. In addition, the authors of the various chapters wish to extend their gratitude to the following people. The frontispiece for chapters 1 and 3 and Figure 1-8 were given by M. A. Mares. For Chapter 4, numerous specialists including D. Duckworth, W. D. Field, D. C. Ferguson, R. W. Hodges, D. R. Smith, E. L. Todd, and D. M. Weisman kindly identified insects collected on *Prosopis*. The authors of Chapter 5 would like to thank G. C. Eickworth, P. D. Hurd, W. E. LaVerge, T. B. Mitchell, J. S. Moure, L. S. Stange, R. R. Snelling, R. W. Thorp, T. H. Timberlake, and D. Urban for help in identification of insects. The illustration, Figure 7-1, was adapted from a photograph lent through the U.S. Fish and Wildlife Service and the photographer. The author of Chapter 8 thanks V. Bohrer, B. L. Fontana, R. I. Ford, V. Jones, D. S. Matson, E. Moser, M. B. Moser, G. P. Nabhan, S. Pablo, A. Russell, and W. H. Wooding for information and criticisms and R. I. Ford for access to the files and collections at the Ethnobotanical Laboratory of the Museum of Anthropology at the University of Michigan. The staff of the Centro Regional del Noreste of the Instituto Nacional de Antropología greatly facilitated investigations in northwestern Mexico. The illustrations used for Figures 8-1 and 8-6 were supplied by the Anthropological Archives of the Smithsonian Institution and Figure 8-3 was photographed by H. Teiwes working with the ethnographic project of B. Doelle. Figure 8-7 was drawn by C. Moser. Figure 10-3 was provided by C. E. Fisher. The many excellent drawings were executed by Biruta Ackerbergs and the graphs and maps by B. B. Simpson and S. Yankowski.

CONTENTS

LIST OF CONTRIBUTORS

Hector L. D'Antoni
  Division Arqueología, Museo de Ciencias Naturales, La Plata, Argentina

Kamalijt Bawa
  Department of Biology, Boston University, Boston, Massachusetts 02115

Arturo Burkart (deceased)
  Instituto de Botanico Darwinion, San Isidro F.C.N.G.M., Bs. As.,
  Argentina

Neil J. Carman
  Department of Botany, University of Texas, Austin, Texas 78712

Rex G. Cates
  Department of Biology, University of New Mexico, Albuquerque, New
  Mexico 87106

Frank A. Enders
  Department of Zoology, University of Texas, Austin, Texas 78712

Richard S. Felger
  Arizona Sonora Desert Museum, P.O. Box 5607, Tucson, Arizona 85703

C. E. Fisher
  Texas Agricultural Experiment Station, Texas A & M University Agricultural Research & Extension Center, R.F.D. No. 3, Lubbock, Texas 79401

Juan H. Hunziker
  Departmento de Genética, Universidad de Buenos Aires, Bs. As., Argentina

Clarence D. Johnson
  Department of Biological Sciences, Northern Arizona University, Flagstaff, Arizona 86001

John M. Kingsolver
Systematic Entomology Division, ARS Entomology Research Division, U.S. National Museum, Washington, D.C. 20560

Michael A. Mares
Department of Biology, University of Pittsburgh, Pittsburgh, Pennsylvania 15260

Andrew R. Moldenke
Board of Studies in Biology, University of California, Santa Cruz, California 95064

Harold A. Mooney
Department of Biological Sciences, Stanford University, Stanford California 94305

Carlos A. Naranjo
Departamento de Ciencias Biológicas, Universidad de Buenos Aires, Bs. As., Argentina

John L. Neff
Division of Biological Sciences, Applied Sciences Building, University of California, Santa Cruz, California 95064

Ramon A. Palacios
Departamento de Botanica, Universidad de Buenos Aires, Bs. As., Argentina

Lydia Poggio
Departamento de Genética, Universidad de Buenos Aires, Bs. As., Argentina

David F. Rhoades
Department of Zoology, University of Washington, Seattle, Washington 98105

Beryl B. Simpson
Department of Botany, Smithsonian Institution, Washington, D.C. 20560

Otto T. Solbrig
Department of Biology and Gray Herbarium, Harvard University, Cambridge, Massachusetts 02138

Stanley R. Swier
Ohio Agricultural Research and Development Center, Wooster, Ohio 44691

Arturo L. Teran
Fundación Miguel Lillo, Miguel Lillo 205, San Miguel de Tucuman, Tucuman, Argentina

ARTIST:

Biruta Akerbergs
Department of Entomology, U.S. National Museum, Washington, D.C. 20560

MESQUITE

FIGURE 1-1. *An aerial view of the Bolsón de Pipanaco near Andalgalá, Catamarca, Argentina. Trees of* Prosopis *(many of the large dark spots) follow the washes winding across the desert.*

Chapter 1

INTRODUCTION

B. B. Simpson, O. T. Solbrig

Spiny plants of *Prosopis* are among the most common plants along washes and canyons of the New World deserts as well as those of northern Africa and the Middle East (Figure 1-2). Since the time of early man, *Prosopis* has played an important role in the lives of peoples inhabiting these arid regions. In ancient cultures of the North and South American deserts, mesquite beans often constituted one of the most important food items. Other parts of the tree provided a wide variety of useful natural products. In South America today, algarrobo fruits are used as animal fodder and *Prosopis* wood is one of the major woods used in the parquet floor industry.

In addition to man, recent studies have shown that *Prosopis* is a dominant part of the natural life in New World desert ecosystems. Its life history and physiology provide prime examples of some of the principal ways in which angiosperms have been selected for survival in hot arid environments. Despite these attributes, however, species of *Prosopis* are now considered to be major rangeland pests in both the southwestern United States and northwestern Argentina because of their increases in population densities. This recent increase in numbers of mesquite trees resulted from the disruption of the pre-1850 ecological balance following the introduction of European agricultural practices.

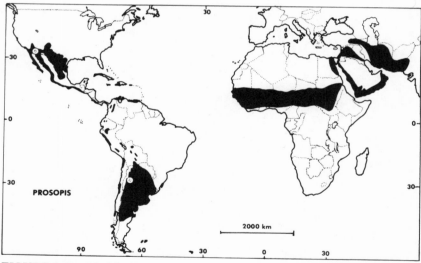

FIGURE 1-2. *Worldwide distribution of* Prosopis. *Three species occur in Asia and adjacent Africa. A fourth is restricted to Africa. Most of the species are indigenous to the New World with the concentration of species (thirty-five) in South America and only nine native to North America. The distributions of both the different sections of the genus and of individual species are given in the Appendix. The dots show the location of the study areas. Only natural distributions are indicated.*

This book brings together studies on the biology and ecology of *Prosopis* species growing in two widely separated (Figure 1-2), but climatically similar (Table 1-1, Figure 1-3) deserts, one in the southwestern United States near Tucson, Arizona, and the other in northwestern Argentina near the town of Andalgalá, Catamarca. These studies were part of the International Biological Program Structure of Ecosystems Project. Its primary objective was to understand the process of convergent evolution in warm desert ecosystems (Orians and Solbrig, 1977). We made no attempt to investigate all of the species of *Prosopis* but tried to obtain a thorough knowledge of the species of mesquite that occur at the two localities chosen for intensive study. We are aware of the enormous amount of research that has been conducted on various species of *Prosopis* but have not attempted to review the literature. Bibliographies of much of the literature are contained in Bogusch (1950) and Schuster (1969). As an introduction, in this chapter we first describe the objectives of the overall project, discuss the two study areas and examine the spectra of plant life at the two localities, and place *Prosopis* in this physical and biological setting. We next examine the major ways in which plants are adapted for survival in desert environments and point out the strategy of *Prosopis*. Finally, we look in detail at the species of *Prosopis* with which we are primarily concerned throughout the following chapters.

THE STRUCTURE OF ECOSYSTEMS PROJECT

The Structure of Ecosystems Project of the International Biological Program was designed specifically to determine to what extent similar physical environments (climate, geology, topography) in widely separated geographical areas has led to the evolution of biotas that play similar roles in the structure and functioning of ecosystems.

As part of this study, two disjunct warm desert regions with similar geological histories (Simpson and Vervoorst, 1977) and climatic regimes (Table 1-1, Figure 1-3) were selected for detailed comparison. Desert ecosystems were investigated not only because matching sites could be found but also because, in many ways, deserts are less complex than more mesic ecosystems. The northern site is the Silver Bell Bajada located at 32° N latitude, 48 km west of Tucson, Arizona, U.S.A. The second primary site in the Bolsón de Pipanaco is at 27° S latitude, 60 km south of Andalgalá, Catamarca, Argentina. Silver Bell is in the northern part of the Sonoran Desert. The Bolsón is in the northern part of the Monte Desert. Over the four-year period of this study (1971-1975), there were numerous investigations of the physical aspects of the two areas and of the biological properties of the component plants and animals.

Much of the actual field work was carried out at localities near the primary sites but not actually at them. Often individual research projects neces-

TABLE 1-1 *Physiography, Geology, and Climate of the Bolsón de Pipanaco and the Silver Bell Bajada*

Characteristic	Silver Bell Bajada	Bolsón de Pipanaco
Latitude	32.4° N	27.8° S
Longitude	61.0° W	66.2° W
Altitude	ca. 739 m	ca. 1072 m
Mean annual temperature	19.5° C	18.6° C
Mean annual precipitation	293 mm	300 mm
Time of permanent emergence	Cretaceous	Cambrian
First traces of climatic aridity	Eocene	Cretaceous
Final formation of arid environment	Plio-Pleistocene	Plio-Pleistocene
Pleistocene effects:		
Glaciers within 200 km	No	Yes
Increase in humidity during cold periods	Yes	Yes
Temperature depression during cold periods	Yes	Yes
Glacial lakes within a few km greater than 100 km^2	No	Yes

Source: Data for the town of Andalgalá taken from twenty year averages (1941–1960). Estadisticas climatologicas. 1941–1950. Servicio Meterol. Nac. Minist. Aeronautica 1958. Publ. 3. p. 32 and 1951–1960, Fuerza Aerea Argentina. Comando de regiones Aereas Serv. Meterol. Nac. 1972. Ser. B. p. 39. Data for the airport at Tucson from L. R. Jurwitz and P. C. Kangieser, 1959. Climate of Arizona. U.S. Dept. of Commerce. Weather Bureau. Climatology of the U.S. No. 60-2. p. 5. Geological data from Simpson and Vervoorst (1977).

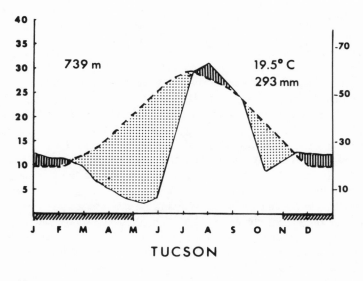

TUCSON

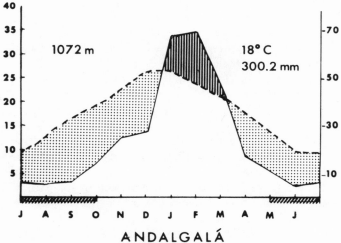

ANDALGALÁ

FIGURE 1-3. *Climate diagrams of Tucson, Arizona, and Andalgalá, Catamarca. The right-hand vertical axis is marked in mm of precipitation and the solid line connects the mean monthly values. The left-hand vertical axis is calibrated in degrees Centigrade. The dashed line connects the monthly means. The stippled areas indicate periods of the year during which evaporation exceeds precipitation and the areas marked with vertical bars show times of excess moisture. The bars below the horizontal axis span months during which freezing temperatures have been recorded. The number in the upper left of the figures is the altitude of the weather station at each site. The numbers in the upper right give the mean annual temperature and mean annual precipitation of the localities. The figures were drawn from the data provided by the sources used in Table 1-1.*

5

sitated a closer proximity to laboratories than afforded by the primary sites (see Figures 1-4 and 1-5).

Geomorphological studies of the two areas showed that both are in basins surrounded by mountains (Figure 1-6) and exhibit characteristic desert landscapes of relatively steep, rocky slopes, gentle gravel pediments and widespread graded deposits of alluvium (Simpson and Vervoorst, 1977). The alluvium and the pediments are commonly called bajadas. The higher, usually steeper portion is the upper bajada and the lower, more gently sloping part, the lower bajada. In valleys between mountain peaks in the Bolsón, alluvium is washed down the valleys and is spread out onto the basin floor forming an alluvial fan. Basins can be closed with an internal drainage, or open with an external drainage system (Figure 1-6). The Bolsón de Pipanaco is a closed basin with a playa or salt flat at the lowest point. Salt accumulation is caused by the surface evaporation of water drained to the playa. Because the Silver Bell Bajada is externally drained, there is no comparable salt flat.

Like all warm desert environments, both of our study areas have limited rainfall, but different parts of the habitat acquire different amounts of effective precipitation. The heaviest rainfall tends to occur on the surrounding mountains. During heavy rains, much water runs off and drains rapidly toward the bottom of the basin. Because of the differences in the quantities of precipitation received and the variable rates at which it is absorbed or flows across the graded slopes, unequal amounts and sizes of rock and soil particles are carried along at different points of the bajadas. On the upper parts of the bajadas, large quantities of fast moving water can carry coarse rocks or even boulders. As the flow dwindles and slows, it begins to drop the heavier particles and carries progressively smaller and smaller gravel. By the time the water reaches the playa, if it has not already ceased to flow, it carries only the finest sand. This process causes the gradation of soil particle size from rock to gravel to sand from mountain slope to bajada to playa. At intervals cutting across all of the landforms are washes or occasional stream beds. Immediately after a rainstorm, many of these washes flood, but they are normally dry for most of the year. Relative to other parts of the desert habitat, however, washes have a greater soil moisture throughout the year because of the concentration of water that soaks into the ground during the rainy season.

The difference in soil surfaces of the habitat and the playa also contributes to the differential absorption of water following a rainstorm. On the upper bajadas where the pediments meet the mountain sides, the combination of steepness and large surfaces of pure rock restricts absorption and promotes high runoff. However, this part of the bajada receives more rainfall (at least in Andalgalá) than lower areas and loses less through evaporation than the flat basin floors. Consequently, the soils of the uppermost bajadas retain relatively more humidity than those in the lower parts of the basin (Mares, 1975). On the lower parts of the bajada, where soils are composed primarily of large irregular rocks and coarse gravel, water is readily absorbed through the many

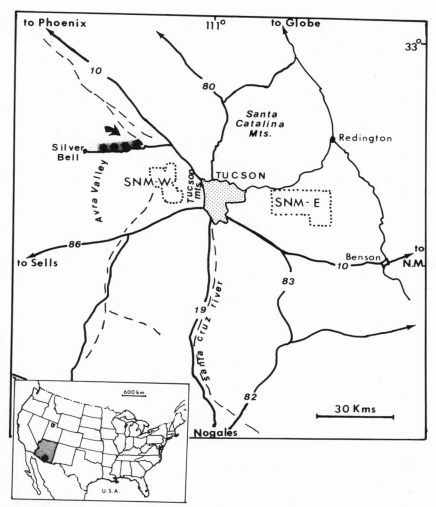

FIGURE 1-4. *The small map shows the location of the study area in Arizona, U.S.A. The large map shows the transect area (arrow and shading with stars) along the road to Silver Bell in the Avra Valley. Phenological studies of* Prosopis velutina *and other desert scrub perennials were made in both Saguaro National Monument West (SNM-W) and Saguaro National Monument East (SNM-E). Ecological studies of* Prosopis *were made at the transect and in Saguaro National Monument West. The junction of Redington is east-north-east of Tucson. Around this area is one of the few remaining natural woodlands of mesquite which contains large old trees (Figure 1-9).*

7

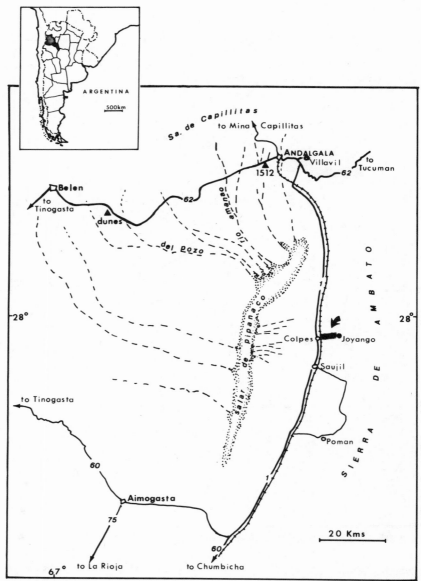

FIGURE 1-5. *The small map indicates the location of the study area in Catamarca, Argentina. The large map gives the details of the Bolsón de Pipanaco where the studies of* Prosopis *were made. The ecological transect (Figure 1-7) is indicated by the arrow, shading and dots, at Joyango. Phenological studies of* Prosopis chilensis, P. flexuosa *and* P. torquata *were made at km 1512 (just west of Andalgalá) and at Villavil (east of Andalgalá). Phenological studies of other desert scrub perennials were made at "the dunes" far west of Andalgalá. Most of the ecological interaction studies were made along the river system at km 1512.*

SILVER BELL

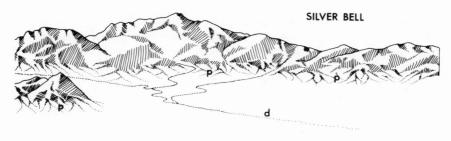

BOLSÓN DE PIPANACO

FIGURE 1-6. *Panoramic views of the valley systems in which the North and South American desert scrub study sites were located. In Arizona (upper figure), the Silver Bell Bajada can be seen to be an open basin with external drainage (d) and incompletely surrounded by low mountains. The ecological transect (Figure 1-7) followed a gradient across upper and lower bajadas on a pediment (p) and continued onto the valley floor. The Bolsón de Pipanaco is a closed basin (lower figure) surrounded by high mountains. The drainage is internal and runoff flows toward the salar, or salt lake, at the bottom of the valley(s). The transect at this site crossed upper and lower bajadas across an alluvial fan (a) and continued across part of the flats.*

interparticular fissures and provides better soil moisture conditions for plant growth than the sandy flats. On the playa, the dry, fine sand retards the rate of water absorption and much of the rainfall runs off before it can percolate. These differences in soil surface and soil moisture across the desert landscape result in a gradient of microhabitats that support varying plant communities.

In order to assess and compare the changes in vegetation in the two study areas across habitat gradients, we made transects from upper bajadas to playas and sampled the perennial vegetation at regular intervals along the transects (Lowe et al., 1977). Detailed records of daily temperature and rainfall during the duration of the project were also made. In both areas, we monitored the phenological patterns of the dominant perennial plants (LeClair et al., 1973a,b; LeClair and Brown, 1974a,b; Yang and Abe, 1973a,b) weekly throughout the same period. Figure 1-7 shows a schematic representation of the transects made at the two areas and indicates the dominant perennial species associated with upper and lower bajadas and flats.

In both areas, the vegetation becomes more diverse, structurally as well as taxonomically, from the flats to the upper bajadas. At Silver Bell, the flats are dominated by the creosotebush, *Larrea tridentata* (Zygophyllaceae), often growing with undershrubs of *Ambrosia dumosa* and *A. deltoidea* (Compositae). On the lower bajadas, there are a few *Opuntia* species (Cactaceae) and scattered individuals of ratany (*Krameria* spp., Krameriaceae) as well as occasional outliers of species that are more common on the upper bajadas. The creosote becomes progressively less important as the habitat changes from the playa to the upper bajada, but a few individuals can always be found. Many of the desert shrubs that are associated with the Sonoran Desert are most common on the upper bajadas. These include the palo verde (*Cercidium microphyllum,* Leguminosae), ironwood (*Olneya tesota,* Leguminosae), saguaro (*Carnegiea gigantea,* Cactaceae) and other cacti, limberbush (*Jatropha cardiophylla,* Euphorbiaceae), ocotillo (*Fouquieria splendens,* Fouquieriaceae), and brittlebush (*Encelia farinosa,* Compositae). Depending on microhabitat conditions, jojoba (*Simmondsia chinensis,* Buxaceae) and fairly paintbrush (*Calliandra eriophylla,* Leguminosae) can be locally abundant. Especially along washes, trees of the velvet mesquite (*Prosopis velutina,* Leguminosae), cat's claw acacia (*Acacia greggii,* Leguminosae) and white thorn acacia (*Acacia constricta,* Leguminosae) are particularly abundant.

During years with good winter rainfall, numerous spring annuals appear. The most common of these are *Eschscholtzia mexicana* (Papaveraceae), *Lesquerella gordoni* (Cruciferae) and species of *Erodium* (Geraniaceae), *Sphaeralcea* (Malvaceae), *Oenothera* (Onagraceae), *Phacelia* (Solanaceae), and *Cryptantha* (Boraginaceae). Toward the end of summer, a second annual bloom occurs. At this time, the most conspicuous species are grasses (*Bouteloua aristidoides* and *B. barbata*), *Tidestromia lanuginosa* (Amaranthaceae), *Allionia incarnata* (Nyctaginaceae), *Kallstroemia grandiflora* (Zygophyllaceae), several species of *Euphorbia* (Euphorbiaceae), and *Pectis papposa* (Compositae).

Near Andalgalá, the flats are also dominated by creosotebush, or jarilla, *Larrea cuneifolia*. There is no understory shrub comparable to either species of *Ambrosia* in Arizona. On the lower bajadas, *Larrea* still dominates but there are scattered individuals of the *retamo* (*Bulnesia retama,* Zygophyllaceae) and *tentitata* (*Prosopis torquata,* Leguminosae). In relatively mesic depressions there are individuals of *pichanilla* (*Cassia aphylla,* Leguminosae) and *brea* (*Cercidium praecox,* Leguminosae). On the upper bajadas, the preeminance of *Larrea cuneifolia* gives way as the *cardon* (*Trichocereus ter-*

scheckii, Cactaceae), various species of *Opuntia* cacti, the *usillo* (*Tricomaria usillo*, Malphigiaceae), *nogal silvestre* (*Jatropha macrocarpa*, Euphorbiaceae), *Prosopis torquata* and *chirqui* (*Mimosa farinosa*, Leguminosae) become common. Small trees of *garabato* (*Acacia furcatispina*, Leguminosae), *tusca* (*Acacia aroma*, Leguminosae), and even *algarrobo amarillo* (*Prosopis flexuosa*) can be found. In localized areas, *jarilla de la puna* (*Zuccagnia punctata*, Leguminosae) is common. Along the broad washes, algarrobos (both *Prosopis chilensis* and *P. flexuosa*), *tala* (*Celtis spinosa*, Ulmaceae), and *tulisquin* (*Grabowskia boerhaaviaefolia*, Solanaceae) form dense stands with shrubs of *atamesqui* (*Atamisquea emarginata*, Capparidaceae) often growing under trees of *Prosopis*.

There is no spring annual bloom in Andalgalá because of the low winter rainfall. However, in summer, numerous species of grasses (e.g., *Aristida adscensiones*, *Bouteloua aristidoides* and *Chloris virgata*) are abundant. Huge areas of the desert landscape can be carpeted with individuals of *Gomphrena martiana* (Amaranthaceae) and *Heliotropium mendocinum* (Boraginaceae) in years with good summer rains. The most common annuals scattered among the perennial shrubs are *Allionia incarnata* and *Boerhaavia coccinea* (both also abundant in the summer bloom at Silver Bell), species of *Portulaca* (Portulacaceae), and the possibly introduced composite, *Verbesina encellioides*.

Physiognomic similarity in plant growth form (clearly visible in Figure 1–7) of the perennial vegetation in regions with similar physical environments has been noted since the time of von Humboldt (1806). As a consequence of this similarity, many ecologists (Schimper, 1903; Mooney and Dunn, 1970) have hypothesized that most other biological properties of such ecosystems should also be similar. If the taxa involved are unrelated, the similarity would be the result of convergent evolution. However, although similar, no two ecosystems are identical and any dissimilarities found can always be ascribed to the inevitable differences in their physical environments. The impossibility of finding perfect environmental matches creates a situation in which it is possible to argue for or against an hypothesis of convergent evolution by placing the emphasis on either the similarities or the differences. Clearly, hypotheses are required that predict realistic levels of similarity or difference. The prediction of such levels of similarity in various components of the two warm desert ecosystem sites and the test of these predictions was one of the primary objectives of the Structure of Ecosystems Project (Orians and Solbrig, 1977, give a summary of these results). One set of these predictions, dealing with the major life forms and growth characteristics of angiosperms expected under desert conditions (Orians and Solbrig, manuscript), suggested that there are a limited number of ways in which plants can effectively deal with xeric environments and that the dominant plant taxa of the two study areas would show convergence in patterns of growth and plant form. The habit and growth of *Prosopis* in the two study areas, as outlined below, represents one of the predicted ways of surviving in the desert. However, in the case of *Prosopis* many similarities found probably

SILVER BELL

UPPER BAJADA   LOWER BAJADA   WASH   FLAT

700

a

BOLSÓN DE PIPANACO

UPPER BAJADA

LOWER BAJADA

WASH

FLAT

PROSOPIS
FRINGE

DUNES

SALAR

1200
1100
1000
900
800

b

FIGURE 1-7. *Schematic representation of the dominant perennial plant species growing on various parts of the desert scrub landscape at the two study sites. The drawings were made from plot sample data along the transects of Lowe et al. (1977 and unpubl.). a. The transect at Silver Bell, Arizona, U.S.A., covers a horizontal distance of 10 km. Species characteristic of parts of the habitat gradient are from left to right: Ef, Ef, Ac, Cg, Oa, Jc, Fs, Cg, Kg, Op, Kg, Cg, Fs, Op, Ce, Cg, Lt, Cg, Jc, Cm, Ag, Of, Kg, Fs, Ad, Op, Cm, Ot, Kg, Cm, Ad, Cm, Oa, Kg, Ot, Kg, Cm, Op, Adu, Lt, Pv, Ac, Adu, Lt, Adu, Lt, Adu, Lt. Symbols: Ac= Acacia constricta. Ad= Ambrosia deltoidea. Adu= Ambrosia dumosa. Ag= Acacia greggii. Ce= Calliandra eriophylla. Cg= Carnegiea gigantea. Cm= Cercidium microphyllum. Ef= Encelia farinosa. Fs= Fouquieria splendens. Jc= Jatropha cardiophylla. Kg= Krameria grayi. Lt= Larrea tridentata. Oa= Opuntia acanthocarpa. Op= Opuntia phaeacantha. Ot= Olneya tesota. Pv= Prosopis velutina. b. Transect at Joyango near Andalgalá, Catamarca, Argentina. Here, the horizontal distance is 20 km. Species characteristic of points along the habitat gradient from left to right continuously across the upper layer of figures: Tu, salar, Br, Lc, Ca, Pf, Ca, Br, Lc, Lc, Lc, Lc, Lc, Tu, Cp, Br, Pc mixed with Pf, Tu, Aa, Pt, Cp, Os, Tt, Tu, Lc, Tt, Af, Tu, Jm, Tu, Jm, Tt, Pf, Ld, Os, Af, Ds, Ds, Af. lower level of figures left to right: Ld, Pf, Af, Ds, Ds. Symbols: Aa= Acacia aroma. Af= Acacia furcatispina. Br= Bulnesia retama. Ca= Cassia aphylla. Cp= Cercidium praecox. Ds= Deuterocohnia schreiteri. Jm= Jatropha macrocarpa. Lc= Larrea cuneifolia. Ld= Larrea divaricata. Os= Opuntia sulphurea. Pc= Prosopis chilensis. Pf= Prosopis flexuosa. Pt= Prosopis torquata. Tt= Trichocereus terscheckii. Tu= Tricomaria usillo.*

must be ascribed to parallel evolution rather than to convergent evolution because the species involved belong to the same genus and are phylogenetically related. Cases which can be explained only by convergence are found, nevertheless, in unrelated organisms that are dependent upon, or are associated with, *Prosopis* in the two deserts.

ADAPTATIONS OF PLANTS TO DESERT ENVIRONMENTS

The intense insolation, atmospheric drought, and high negative soil-water potentials of deserts pose severe constraints on the growth and maintenance of angiosperms. To overcome these strictures, plants must possess special features, especially in the morphology and physiology of their leaves, roots, and conducting systems. In order to absorb water from the soil, a plant exerts a "suction" force—that is, it produces a water potential within the root lower than the force with which water molecules adhere to the surrounding soil. The water potential within the root is negative relative to the soil and water moves into the root hairs. As a necessary part of the process of photosynthesis, a plant has to take in carbon dioxide from the air. As the carbon dioxide is used by photosynthetic (usually leaf) tissue, its concentration in the air spaces within the tissue drops to a very low level causing a gradient from inside the leaf to the outside air. Carbon dioxide consequently continues to diffuse into tissues through openings called stomates. While the stomates are open, and carbon dioxide is moving into the leaf, water vapor which has a higher concentration within the leaf than without, inevitably diffuses out. Water loss by this process, known as transpiration, can be sizable. Water is constantly lost from leaves and must be replaced by water that has been absorbed by the roots and conducted up through the plant. As the soil around the root becomes drier, it becomes increasingly difficult for the plant to absorb water. What little moisture there is in the desert soil is held with increasingly stronger cohesive forces and the root has to exert a constantly greater negative water potential. However, most plant species can withstand only low to moderate negative water potentials (less than -20 bar). When a plant has extracted all the water it is physiologically capable of extracting (or reached its maximum negative water potential) it is forced to close its stomates to stop water loss by transpiration or suffer irreversible damage by permanent wilting of its leaves. Closing the stomates means that photosynthesis must stop because carbon dioxide is prevented from diffusing into the leaf. Furthermore, most species cannot prevent the loss of small amounts of water through the leaf epidermis even after the stomates are closed. As a result, permanent wilting is inevitable unless soil water is replenished.

In deserts, potential evaporation greatly exceeds rainfall for most of the year. Soils consequently become exceedingly dry, especially in the uppermost layers. Under such arid conditions, mesophytic plants soon wilt and die. The relatively small number of plants that survive in the desert do so because they have found solutions to the problem of severe water loss. Solutions differ, but several major syndromes of drought avoidance or endurance have been evolved in most warm desert regions.

One method of escaping times of drought, exemplified by the colorful desert annuals and ephemerals, is to grow only during that part of the year (or during years) when there is sufficient soil moisture for growth. Seeds of these plants germinate with the first substantial rains and the plants complete their entire life cycle in the brief one to three months before the desert becomes too dry again for growth. Woody species that leaf when it is rainy and drop their leaves during dry periods have similar drought avoiding systems.

Another method of enduring, rather than avoiding, drought is that found in cacti and other succulents. During the rainy season, these plants absorb and store large quantities of water in their tissues. The amount they can store is, however, only a fraction of what plants normally lose through transpiration. Succulents must therefore still possess adaptations to minimize water loss. For example, they open their stomates on the stems or fleshy leaves at night when ambient temperatures are low and transpirational loss is reduced. This behavior pattern requires a specialized photosynthesis system known as CAM (crassulacean acid metabolism) (Mooney, 1972). Succulents also possess thick cuticles and other morphological adaptations that reduce heat loads and loss of water through the epidermis.

A third way of coping with desertic climates is to have cells capable of withstanding high negative water potentials. Such plants can extract water from soils too arid for most plant species and can extend their growing season through the dry periods of the year. The few species of halophytes and evergreen shrubs that have been able to evolve such physiological systems are characteristic of deserts and semi-deserts and constitute the primary examples of true xerophytes (meaning dry loving). The most widespread xerophytic shrubs in the warm deserts of the New World are different species of creosotebush, or *Larrea*. These shrubs, although capable of an extended season of productivity, are nevertheless affected by periods of prolonged drought and are often forced to close their stomates. They also possess a number of specialized adaptations such as small leaves with thick outer walls of the epidermal cell layers (Hull et al., 1971), sunken stomates, specialized conducting systems, and extensive root networks that minimize the effects of dry air and soils. As a part of the Structure of Ecosystems Project, the adaptations of *Larrea* to desert conditions were studied in detail (Mabry et al., 1977). For purposes of our study, *Larrea* constitutes the outstanding example of a xerophytic drought enduring species (Table 1-2).

In direct contrast to *Larrea* and exemplifying a final way in which plants survive in the desert are woody perennials that grow in and exploit water rich microhabitats (Figure 1-7). These species called phreatophytes from *phreatos*, well, and *pheitin*, to grow (in), have exceedingly long tap roots as well as more shallow lateral root systems that allow them to tap underground water below the surface of the soil when the surficial supply is depleted. The most

TABLE 1-2 *Comparative Features of a Warm Desert Woody Xerophyte and Phreatophyte*

	XEROPHYTE *LARREA*	PHREATOPHYTE *PROSOPIS*
Adult habit	shrub	tree
Leaves	evergreen, xerophytic characteristics	deciduous, mesophytic characteristics
Root system	primarily in upper 4 meters of the soil	upper layer and below 4 meters in the soil
Sexual maturity	can flower after four years	flowers after three years
Flowering	patchy in space	synchronous over large areas
Breeding system	facultative outbreeder	obligate outbreeder
Pollination	primarily bees	primarily bees
Fruiting	produces an unpredictable, undependable crop	produces a comparatively "non-failing" crop
Fruit	dry, dehiscent	fleshy, indehiscent
Principal dispersal agents	wind, water	animals
Seed	moderate size, rapid germination with moisture	large size, long dormancy unless freed from endocarp, germination aided by scarification
Germination and seedling initiation	small endosperm does not allow extensive initial seedling growth	copious endosperm allows rapid growth with pronounced root development

common phreatophytic species in the study areas are members of the genus *Prosopis*. The contrast between *Larrea*, a typical xerophyte, and *Prosopis*, the principal phreatophyte in the two desert systems, is shown in Table 1-2. Usually, phreatophytes are restricted in true deserts to wash edges where they can reach underground water sources, or to low areas of high water holding capacity. In less arid regions where rainfall is relatively high or local factors such as elevation (Figure 1-7) or exposure increase the amount of superficial available moisture, individuals of *Prosopis* do not necessarily behave like phreatophytes. Despite their specialized root systems, phreatophytes are still exposed to high solar insolation and hot dry air. They consequently have specializations in morphology, physiology, and phenology that tend to reduce high transpiration rates. These adaptations are discussed in detail in Chapter 2.

THE MESQUITES AND ALGARROBOS OF
SILVER BELL AND ANDALGALÁ

In North America, species of *Prosopis* with straight fruits are commonly grouped together under the name mesquite, derived from an Indian (Nahuatl) name, *mesquitl* (Robelo, 1948). Species with tightly coiled fruits are called screwbeans or *tortillos* in North America and *mastuerzos* in South America. The name *algarrobo* is applied to most of the South American species with straight or arched fruits. The common name algarrobo was first used for a species of *Prosopis* by the Spaniards because they confused *Prosopis* in South America with the carob tree, *Ceratonia siliqua* (Leguminosae) known as algarrobo in Spain. As indicated by the type and shape of its fruits, *Prosopis* belongs to the Leguminosae or bean family. On a worldwide basis, the genus includes forty-four species considered to form five natural groups or sections. Forty of the forty-four species occur in the New World. The remaining four species are restricted to Africa or the Middle East. The Appendix describes the taxonomic circumscription of the genus and provides an annotated key to all of the species of *Prosopis*. Synonyms of the various species are given in the Index to Scientific Names.

At the Silver Bell site, there is only one species of mesquite, *Prosopis velutina*, the velvet mesquite. In this area under natural conditions, mesquite grows in low areas of playas and along wash systems (Figure 1-8) and plants are large trees (Figure 1-9). In areas that are grazed and trampled, plants become shrub-like. Figures 1-9 and 1-10 contrast the two growth forms, both of which can be found at or near Silver Bell. Morphologically, this species can be distinguished from its nearest neighbor and relative, *Prosopis glandulosa*, the honey mesquite, by the velvety texture of its leaves and small sized leaflets (Figure 1-11). The pods of the velvet mesquite are quite straight and can vary in color from pure yellow to mottled red or black (Figure 1-11, Figure 6-1). The modern geographical distributions of both the velvet mesquite and the honey mesquite are shown in Figure 1-12.

FIGURE 1-8. *Upper photograph. The low lying areas of the Silver Bell Bajada from a hilltop vantage point. A line of trees, principally* Prosopis, *outlines the course of a stream bed along the valley floor. Lower photograph. A line of trees, primarily species of* Prosopis, *grows along a wash system in the Bolsón. The ecological transect ended at the road (appearing as a white line crossing the picture) but the wash trees continue to follow the water courses to the edge of the salar.*

18

FIGURE 1-9. *An old tree of* Prosopis velutina, *the velvet mesquite, found at Redington, north of Tucson, Arizona (Figure 1-4). This locality was considered to be one of the few areas where an undisturbed* Prosopis *woodland could be seen in 1905 (Meinzer, 1927). Old, single-stemmed individuals such as this are now exceedingly rare because of cutting, clearing, and grazing. Even at Redington cutting has reduced the number of such trees, but several still remain.*

19

FIGURE 1-10. *The characteristic multi-stemmed appearance of most modern* Prosopis velutina *trees, including those at the Silver Bell study site, has been caused by man's disturbance (Chapter 9). In addition to a change in the predominant growth form, human influence has led to increased* Prosopis *population densities.*

In the Bolsón de Pipanaco, there are two dominant species of algarrobos, *Prosopis chilensis* and *P. flexuosa.* Plants of both species are normally large trees (Figure 1-13) that grow along the edges of broad wash systems characteristic of the Bolsón (Figure 1-1). In the lowermost part of the basin where the water table almost reaches the surface of the soil but soil salinity is sufficiently low, woodlands of *P. flexuosa* form a ring around the inhospitable salt flat (Figure 1-7). The most distinctive of these two species is *Prosopis chilensis* (locally called *algarrobo de Chile*) which has very large leaves and leaflets and curved, flattened, yellow pods (Figure 1-11). This species was emphasized in the studies at Andalgalá, but numerous projects also investigated aspects of the biology of the sympatric *P. flexuosa.* This second species, commonly called *algarrobo amarillo,* has very small leaflets that are widely spaced on the rachis and straight, usually mottled fruits (Figure 1-11). The present distributions of these two species are shown in Figure 1-14. In contrast to *P. chilensis, P. flexuosa* occasionally grows on the mountain slopes and uppermost bajadas as well as along washes and around the salar (Figure 1-7).

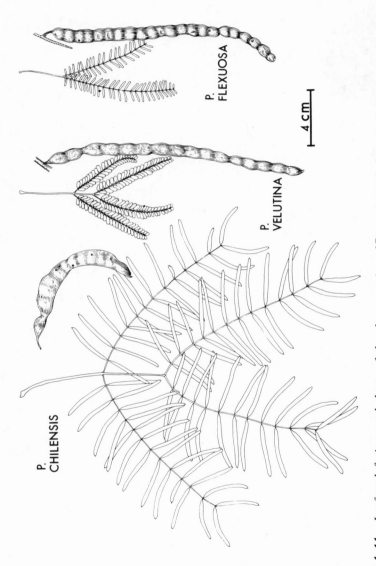

FIGURE 1-11. *Leaf and fruit morphology of the three species of Prosopis studied in detail. The most distinctive species, P. chilensis* (algarrobo de Chile or algarrobo blanco) *occurs with* P. flexuosa (algarrobo amarillo or algarrobo negro) *in the Bolsón de Pipanaco. The only species of Prosopis at Silver Bell is the velvet mesquite,* P. velutina. *Morphological and chemical characters suggest the* P. velutina *and* P. flexuosa *are more closely related to one another than either is to* P. chilensis.

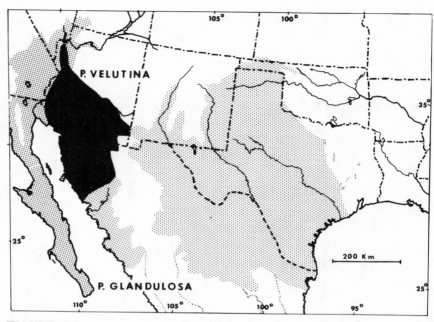

FIGURE 1-12. *The modern distribution of* Prosopis velutina, *the velvet mesquite and its close relative* P. glandulosa, *the honey mesquite, in the southwestern United States.*

FIGURE 1-13. *An old tree of* Prosopis flexuosa, *algarrobo amarillo, in a* Prosopis *woods, or bosque, near Molinos, a town north of Andalgalá. Large individuals with single trunks are the most common growth form of* P. chilensis *and* P. flexuosa *along washes near Andalgalá and multistemmed small trees, common in Arizona, are rare. The browse line is produced by goats. In the southwestern United States, cattle rather than goats are the most common herd animals. Since cattle do not significantly browse mesquite, browse lines are uncommon at the northern site (compare Figure 1-9).*

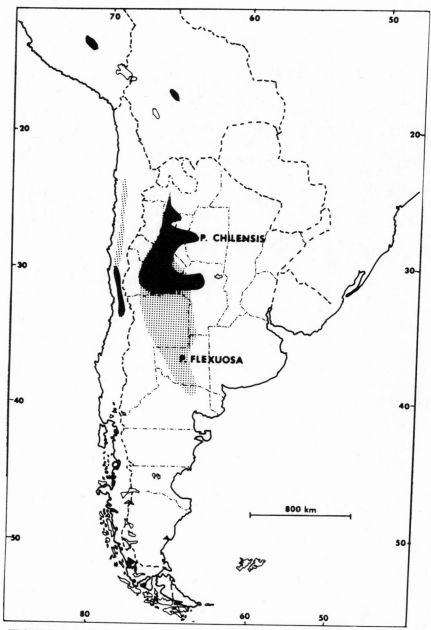

FIGURE 1-14. *The modern distribution of* Prosopis chilensis, algarrobo de Chile *and* P. flexuosa, algarrobo amarillo, *in Argentina and adjacent Chile.*

There are no true screwbeans (section *Strombocarpa*) at Silver Bell, but one species belonging to this section, *Prosopis torquata,* occurs on bajadas in the Bolsón de Pipanaco (Figure 1-7). This species is not a phreatophyte and we will, in general, deal with it only peripherally and for comparative purposes in the following chapters.

SUMMARY

Various adaptive strategies have been evolved for the avoidance or endurance of prolonged periods of drought in the warm desert regions of the world. One of these strategies is that of a phreatophyte, a plant that taps underground water unattainable by most plant species. In the New World deserts, species of *Prosopis* are the dominant native phreatophytes. As part of a project investigating convergent evolution in two disjunct desert ecosystems, we carried out an intensive study of the biology and ecological relationships of *Prosopis* found in the two desert sites chosen for study. These species, *Prosopis velutina* at the Silver Bell, Arizona, U.S.A., site and *P. chilensis* and *P. flexuosa* at the Bolsón de Pipanaco, Catamarca, Argentina, site are important natural components of their respective ecosystems and exhibit similar adaptations to desert life. Chapters 2 and 3 analyze morphological, physiological, and phenological adaptations of the three species. Chapters 4 to 7 treat vegetative, flowering, and fruiting parts of the life cycle as food resources or habitat components of other plants and animals. Many of the relationships between *Prosopis* and other organisms involve mechanisms such as complex toxins for repelling or attracting various animals. Some of the relationships with other organisms are obligatory, others are casual or even secondary. Throughout these chapters we point out those features that represent parallel adaptations and indicate instances of convergent evolution involving organisms associated with *Prosopis*. Finally, Chapters 8 to 10 discuss the widespread utilization of *Prosopis* by native peoples in North and South America and examine its present relationship to the current European-based agricultural society.

Chapter 2

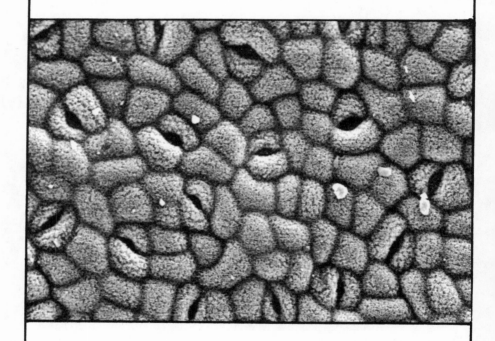

PHENOLOGY, MORPHOLOGY, PHYSIOLOGY

H. A. Mooney

B. B. Simpson

O. T. Solbrig

The principal adaptive problem of woody plants in the desert is no different from that of all land plants—the acquisition and retention of moisture. In deserts, however, the problem is magnified since environmental water is scarce and the competition for it unusually great. In this chapter we ask, how have individuals of *Prosopis* met the challenge of obtaining and utilizing water in desert environments? In answering this question, we take an overall view of *Prosopis,* in particular those species studied at Andalgalá and at Silver Bell. We also include data from studies on *P. glandulosa* from areas adjoining the northern site. First we examine the phenological patterns of the adult plant and the data on the functional significance of its habit, anatomy, and leaf morphology. Then we describe what is known about the water economy and processes of photosynthesis of *Prosopis.* Finally, we briefly outline the adaptations of critical parts of the life cycle, those involving fruit, seed, and seedling. Success in these stages ultimately leads to the dispersal, germination, and establishment of mesquite in desert environments. We are primarily concerned in this chapter with how kinds of adaptations are correlated with the demands imposed by the physical environment. The following chapter deals with sources of variation within and between populations, and Chapters 4 to 7 discuss in detail the relationships of *Prosopis* and the biological components of the environment.

PHENOLOGY*

During the two years of our study, detailed measurements were made of the phenology of *Prosopis velutina* in Arizona near Silver Bell and of *P. chilensis, P. flexuosa* and, to a lesser extent, *P. torquata* in Argentina near Andalgalá (LeClaire et al., 1973a,b; LeClaire and Brown, 1974a,b; Yang and Abe, 1973a,b). These studies dealt with the temporal patterns of leaf formation, flowering, and fruiting.

In Argentina, both *Prosopis chilensis* and *P. flexuosa* start to produce leaves at the beginning of spring (September, with an average temperature of $\pm16°C$) as the weather begins to get warmer and both stay in leaf until fall (April). A similar, although more variable pattern was found in *P. velutina.* The greater variability of the pattern in the northern area can be attributed, in part, to the bimodal distribution of the rainfall (Figure 1-3; Bailey, 1977).

*B. B. Simpson and O. T. Solbrig.

FIGURE 2-1. *The upper leaf epidermis of a leaflet of* Prosopis velutina *from Silver Bell, Arizona, magnified 400 times and viewed with a scanning electron microscope. The granular appearance of the surface of the cells is produced by the accumulation of wax crystals. The stomates are not sunken nor protected by trichomes and thus appear more like those of mesophytes than true xerophytes.*

Initiation of leafing appears to be rather independent of rainfall in all of the species of section *Algarobia* with which we were concerned. The majority of the rainfall in Argentina (Figure 1-3) falls in January and February, well after the leaves are fully mature. Winter rains are essentially negligible. Although winter rainfall can be substantial in Arizona at Silver Bell, *P. velutina* begins to leaf after the rainy season has ended. In Argentina, both *P. chilensis* and *P. flexuosa* invariably lose their leaves during the fall. In Arizona, *P. velutina* usually sheds its leaves at the end of the summer. However, during years with warm winters, trees can retain their leaves throughout the winter, but shed them in the spring when the new leaves are produced.

Populations of *Prosopis glandulosa* that have been studied in detail (Mc-Millan and Peacock, 1964; Peacock and McMillan, 1965) show genetically-based variation in the timing of leaf bud burst and time of leaf drop. In their studies, plants originating from widely separated geographical areas and with differing times of leaf initiation and drop were grown under uniform conditions at Austin, Texas. Individuals retained the timing of bud burst and leaf drop characteristic of their original habitat. Plants from populations in northern latitudes showed late spring bud activity, active growth under long days, and early fall dormancy under short day conditions. Plants from southern populations in Mexico retained early spring bud activity at Austin and showed little correlation with photoperiod. It appears that in the southern populations, temperature, rather than day length, induced fall dormancy. However, as indicated below, in other populations of *P. glandulosa,* a combination of soil moisture, prior drought, and temperature apparently determine the timing of leaf drop. We do not know if *P. velutina, P. chilensis,* and *P. flexuosa* exhibit ecotypic variation in physiological mechanisms for the avoidance of bud freezing across latitudinal gradients.

Flowering at the Arizona and Argentina study sites occurs shortly after leafing begins (Figures 5-2, 5-3). Both *Prosopis chilensis* and *P. flexuosa* (although only the former is shown in Figure 5-2) bloom in the spring around the last week in October at Andalgalá. The first flower bud formation was observed about a month and a half before actual flowering began. Blooming time in these species is relatively constant from year to year and lasts two weeks. While in bloom, the large trees are covered with several thousand, five- to ten-cm-long cream-colored inflorescences similar to those of *P. velutina* (Figures 5-1, 5-4). However, fewer than 3 percent of the millions of flowers produced initiate fruit development and only about one-half to one-third of these subsequently produce fruits (Solbrig and Cantino, 1975). Fruits take approximately three months to mature.

The blooming pattern of *Prosopis velutina* is more irregular than that of the South American algarrobos (Figure 5-2). However, the overall pattern of a predominant spring bloom, a low percentage of the flowers ultimately producing fruits and a prolonged period of fruit maturation is similar to that observed in the species at the Argentine site although the period of time for fruits to mature was longer in *P. velutina* than in either *P. chilensis* or *P. flexuosa*. According to Fisher (personal communication), *P. glandulosa* in western Texas can bloom up to four times per year. In this species, flower production varies with the amount of moisture available. When there is low soil moisture, there is heavy flowering followed by abundant fruiting. If there is high soil moisture, flowering appears to be suppressed and fruit production is low. Rainfall during the flowering period results in low fruit production. A similar pattern was noted in some species of *Prosopis* in eastern Argentina by Burkart (1937).

Our phenological observations indicate that flowering and fruiting in *Prosopis velutina* also varies from year to year and from tree to tree within a population during the same year. As described in Chapter 5, some of the intrapopulational variation in fruiting can be ascribed to differences in attractiveness of inflorescences to insect visitors during a particular flowering season. In addition, fluctuations in the populations of sucking insects (see Chapter 7) probably play a significant role in determining the number of incipient fruits that abort in different years.

HABIT, LEAF AND STEM ANATOMY*

Under undisturbed conditions, the natural habit of *Prosopis chilensis, P. flexuosa* and *P. veultina* are similar. All are generally single stemmed due to the fact that species of the section *Algarobia* show a strong tendency for apical dominance (Meyer et al., 1971) and consequently for a well-developed crown (Figures 1-10, 1-14). At the southern study area, individuals of *P. chilensis* and *P. flexuosa* were characteristically large trees with well-developed trunks and thick crowns, but plants of *P. velutina* at Silver Bell were commonly multi-stemmed and shrubby (Figure 1-10). According to Fisher (see Chapter 9) the formation of a branched trunk results from the destruction of the main shoot or apical meristem by animals or physical processes such as freezing.

Leaves of all three taxa, like all of the species of the genus, are doubly compound (Figure 3-3). Only the cotyledons and the first, or occasionally the second and/or third, pair of leaves are simple or singly compound. Most species have two secondary rachises, but sometimes one or three pair may be present. The margins of the leaves are entire and the leaf surface glabrous in both *P. chilensis* and *P. flexuosa*. The velvet mesquite, *P. velutina*, has single-celled glandular trichomes and wax crystals on the leaf surfaces producing a velvety texture (Figure 2-1) from which the species derives its common name. These three species have stomates on both sides of the leaf, as does *P. glandulosa* (Meyer et al., 1971). As in this neighboring species, the three species of

*O. T. Solbrig.

section *Algarobia* at our study sites have about twice as many stomates on the upper surface of the leaf as on the lower, although the number per unit area varies from tree to tree and within a tree depending upon the position of the leaf. The stomates are not sunken (Figures 2-1, 2-2) or covered by trichomes as is often common in the leaves of xerophytic shrubs (Pyykkö, 1966; Fahn, 1974). However, the surface is covered with wax which often forms characteristic patterns on the leaf surface (Bleckmann and Hull, 1975; Figures 2-1, 2-2). Stomates are also found on the rachises of the leaves. At the base of each leaflet, leaf, and petiole are pulvini that allow the pinnae and leaflets to droop and/or to close. In all the species in section *Algarobia* at the study sites, the leaflets close somewhat at night and the leaves tend to droop partially during the hottest part of the day in summer. The leaves of these three species also have extra-floral nectaries, the function of which is still unknown. Extra-floral nectaries are common in most species of *Prosopis* and in many other mimosoid legumes.

After the first year of growth, the lateral buds on a branch produce leaves, stems, spines, and inflorescences. Each branch shoot in the species of section *Algarobia* produces a series of nodes and internodes. Each node produces one leaf plus two or three buds with all of the leaves arranged in a two-thirds spiral, alternate phylotaxy. After growth has stopped for the year, the terminal bud aborts. The following year a bud from one of the first three internodes becomes the pseudoterminal bud and produces the growth for that year. This pattern gives the branches a slight zigzag appearance.

Above each leaf are axillary buds that may give rise to spines. The spines in the species of *Prosopis* section *Algarobia* are, therefore, modified stems (Figure A-1). Thorns are produced only on new wood and are continuously being buried by secondary xylem (Meyer et al., 1971). After about three years, the original leaf scar is completely buried (Meyer et al., 1971).

The wood of all our species is made up of fibers, parenchyma tissues, and vessels. The wood is ring porous with the vessels scattered in bundles of two to seven. Based on a study of *P. glandulosa* (Meyer et al., 1971), the vessel diameter appears to be large, as would be expected of a relatively mesophytic species. All of the species in our study areas, as well as all of the other species of section *Algarobia* studied, have gum in the heartwood that eventually occupies most of the center of the stem. In older, or wounded trees, this gum can be seen as a thick black exudate running down stems or the tree trunk.

WATER ECONOMY*

Data on the complete water economy of any single species of *Prosopis* are lacking, but enough facts have now been assembled to piece together the pattern for the northern species, *P. velutina* and *P. glandulosa*. Unfortunately,

*H. A. Mooney.

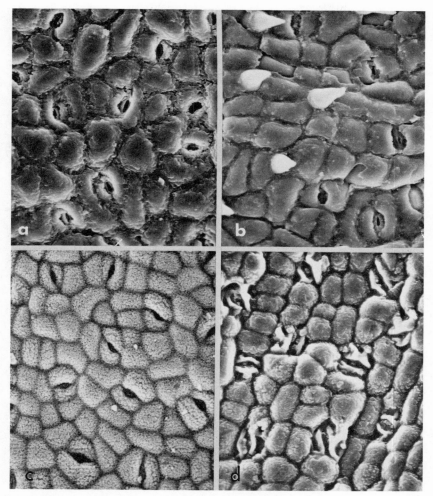

FIGURE 2-2. *Leaf epidermal patterns of* Prosopis chilensis, P. flexuosa, P. velutina *and* P. torquata. *The surface of the leaves of all these species is covered by a coating of wax, but in the southern hemisphere taxa, the wax forms smooth sheets. The stomates of the screwbean,* P. torquata, *the most xeric adapted of the four taxa pictured, are more sunken than those of the three phreatophytic species of section* Algarobia.

equipment necessary for such studies (Figure 2-3) was not available in Argentina and comparable data are thus not available.

Unappreciated until recently because of the mesophytic appearance of the external morphology of mesquite, the physiological capabilities of *Prosopis* for coping with drought are complex and, on a short-term basis, can rival those of desert xerophytes. Adaptations for acquiring and retaining moisture

FIGURE 2-3. *A leaf of* Prosopis glandulosa *inside a* cuvette, *a chamber used for the measurement of photosynthesis and respiration. Both temperature and humidity can be held constant within the chamber as air is passed over the leaf.*

involve the root system, the leaf morphology, and the physiological tolerances of the plant.

As has been known for some time, the potential water reservoir available to *Prosopis* is large relative to that of other desert shrubs. Trees have roots deep enough (in some cases) to tap ground water as well as a pronounced lateral root system. Phillips (1963) reported that roots of *P. velutina* south of Tucson were excavated at a depth of almost 50 m, and Fisher et al. (1959) cite examples of trees of *P. glandulosa* with lateral roots extending over 18 m from the tree. It is obvious that water in the upper soil horizon as well as that in deeper layers is utilized whenever it is available (Haas and Dodd, 1972) and apparently is used preferentially above deeper sources (Easter and Sosebee, 1975). Like desert xerophytes, *Prosopis* can acquire soil water which is held with rather high matric forces. Haas and Dodd (1972) recorded reductions in the soil water under plants of *P. glandulosa* as low as -15 bar at depths to at least 150 cm.

In addition to its capability for absorbing water held under high forces, we can ask if mesquite has specialized adaptations promoting efficient utilization and retention of the water it acquires. Indeed, data show that *Prosopis glandulosa* actively carries on photosynthesis at xylem water potentials of less than -40 bar (Strain, 1970) emphasizing that *Prosopis* has xerophytic characteristics also in terms of water utilization. Nevertheless, even though *Prosopis* may have comparable water potentials at dawn to those of *Larrea tridentata* (Figure 2-4), the classic xerophyte, when they occur together in the same habitat, *Larrea* will be under greater moisture stress by midday (Strain, 1970). The similarity of the dawn water potentials is interesting considering that *Prosopis* presumably has the ability to tap different water sources from *Larrea*. The difference in the midday xylem water potentials could come about in a number of ways, but it is apparent that one of the most likely is that *Prosopis* is able to use water more conservatively than *Larrea*, particularly during periods of high evaporative demand. Wendt et al. (1968) postulated the presence of mechanisms to reduce transpiration in *P. glandulosa* at high vapor pressure deficits and Easter and Sosebee (1975) further noted that as soil moisture diminished, a lower vapor pressure deficit produced a decrease in transpiration. In other words, if the evaporative demand of the air exceeds a certain level, *Prosopis* essentially becomes decoupled from the atmospheric environment (transpiration is generally linearly related to the vapor pressure deficit assuming equal stomatal openings). The point at which this decoupling occurs appears to shift with the amount of available soil moisture.

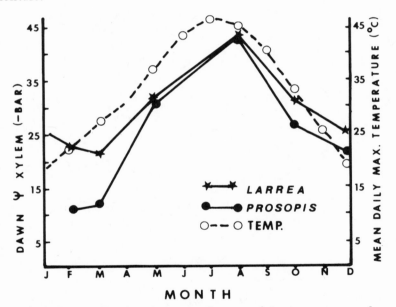

MONTH

FIGURE 2-4. *Comparative seasonal course of air temperatures and of dawn water potentials of* Larrea tridentata *and* Prosopis glandulosa *in Death Valley, California (data from Mooney et al., 1975). Even though the two taxa presumably tap different water sources, they show similar dawn water potentials throughout much of the year. The lower values for* Prosopis *during the winter are a reflection of the deciduous habit of mesquite.*

The mechanism behind these observations involves a direct response of the stomates to vapor pressure deficits (VPD). As the VPD increases to a point at which water loss would be so great as to result in a very high water stress, the stomates simply close (Figure 2-5). This stomatal closure results, of course, in a concommitant reduction in photosynthesis since it blocks the intake of carbon dioxide from the atmosphere into the leaf as well as the escape of water. The midday closure of the plant stomates in desert environments can enhance water use efficiency considerably because it increases the ratio of carbon fixation to water loss. Carbon is thus gained only during periods of low vapor pressure deficit which are also the periods of lowest potential water loss.

In order to use effectively a mechanism such as midday stomatal closure in a high radiation desert environment, it would appear that small leaf size, as is found in *Prosopis,* would be mandatory because the transfer of energy through transpiration would be blocked and small leaves or leaflets have a much higher potential for conductive transfer of heat energy than large leaves.

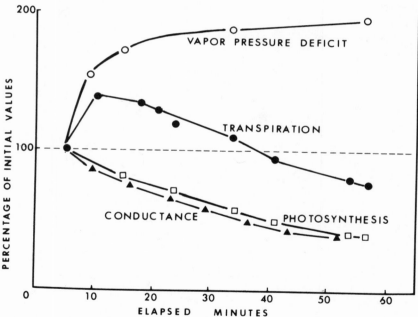

FIGURE 2-5. *Response of* Prosopis glandulosa *to increased vapor pressure deficits. The measurements were made with a plant growing under natural conditions in Death Valley, California, in May 1976. Leaf temperature was held constant at 35°C and irradiance at 170 nEinsteins per cm² per second (400–700 nm). The stomatal closure (decreased conductance) as vapor pressure increased resulted in a decreased water loss and photosynthetic rate (unpublished results, Mooney, Björkman, and Ehleringer).*

(See Chapter 3 for a discussion of the changes in leaf and leaflet size across temperature and moisture gradients.) The water potentials taken at dawn indicate that this is the time of day of minimum water stress for mesquite plants.

We have less knowledge of the physiological ecology of the southern hemisphere *Prosopis* species although what is known of one of them, *P. tamarugo,* indicates that it is one of the world's unique plants. The tamarugo grows in the Atacama Desert in northern Chile under essentially rainless conditions. This species does not have roots that tap the water table. Instead, individuals appear to derive all of their moisture from the atmosphere (Went, 1975). Nevertheless, the soil surrounding the roots is usually moist and the plants do not experience low water potentials (Dunn, personal communication). After Sudzuki (1969) demonstrated that atmospheric moisture was transported to the soil zone, Went (1975) postulated that water is taken up by the leaves at night through dew formation on the leaves. Transport of moisture downward during the night could be the result of the low water potentials of the soil caused by its high osmotic concentration. In this desert, high osmotic differentials are caused by the fact that the plants grow in a thick salt crust. Clearly, further study is needed on this unique transport mechanism.

Experiments on the transpiration of *Prosopis tamarguo* have not shown the response to vapor pressure deficit described above for *P. glandulosa* or *P. velutina.* Whether this difference is caused by a lack of such a response in *P. tamarugo* or whether the experimental conditions were such that a response was not induced (lower vapor pressure deficits were used in the experiments with *P. tamarugo* than with *P. glandulosa*) is not known. It is possible that there is a differential response since the Atacama Desert is a relatively cool (adjacent to a cold coastal upwelling) desert and high vapor pressure deficits do not occur regularly. Although *Prosopis chilensis* grows in the Atacama Desert as well as in the northern Monte Desert (Figure 1-14), it occurs where there is a high water table (Went, 1975) and in habitats that are less extreme than those inhabitated by *P. tamarugo* and more like those of *P. velutina* and *P. glandulosa* in North America.

PHOTOSYNTHETIC CAPACITY*

Detailed information on the seasonal photosynthetic capacity of the various species of *Prosopis* is still lacking. The preliminary information we have gathered on *Prosopis glandulosa* indicates that under non-limiting conditions, it has a carbon gaining capacity which exceeds that of *Larrea tridentata* (Figure 2-6). At an internal CO_2 concentration of 250 μbars, typical of plants in nature, the rate of carbon gain of *Prosopis* exceeds that of *Larrea* by a third. However, measurements on plants of *L. tridentata* and *P. glandulosa*

*H. A. Mooney.

growing side by side in the field in May in Death Valley, California indicate that this potential difference between the two species may not be realized under natural conditions. Both taxa (Figure 2-7) have similar photosynthetic rates at the highest irradiance values measured (approximately those of midday sun). It is possible that at earlier times of the year when temperatures are cooler and the *Prosopis* leaves younger, a differential in favor of mesquite may be realized. It should be noted that the measured photosynthetic values

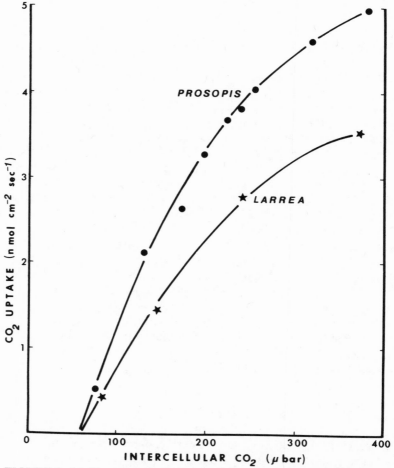

FIGURE 2-6. *Photosynthetic response to carbon dioxide of* Prosopis glandulosa *and* Larrea tridentata. *Plants of both species were grown in a phytocell with light, nutrients, and water unlimited and a daytime temperature of 35°C. Night temperatures were held constant at 25°C. Photosynthetic rates were measured at 30°C and at an irradiance of 160 nEinstein per cm² per second (400–700 nm) (Mooney, unpublished data). Under these unlimited environmental conditions,* Prosopis *has a photosynthetic capacity exceeding that of* Larrea.

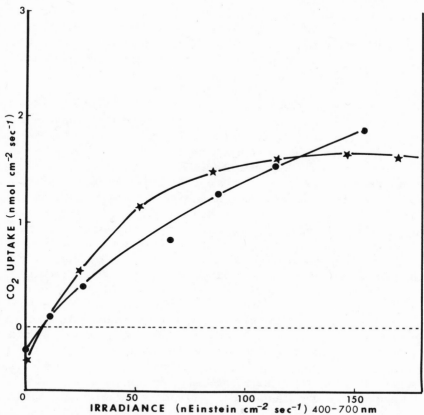

FIGURE 2-7. *Photosynthetic light responses of* Larrea tridentata *and* Prosopis glandulosa *in May (1976) growing under natural conditions in Death Valley, California (unpublished data, Mooney, Björkman, and Ehleringer). Symbols as in Figure 2-6.*

for *Prosopis* (Figure 2-7), two nanomoles of carbon dioxide per square centimeter per second (equivalent to over 30 mg CO_2 per dm^2 per hour), are in excess of values given for the sun leaves of either deciduous broad-leaved trees in general or evergreen broad-leaved trees (Šesták et al., 1971). Values for the former average about 10-25 mg CO_2 per dm^2 per hour and for the latter, 10-16 mg CO_2 per dm^2 per hour.

Undoubtedly, the photosynthetic rates of *Prosopis* change considerably during the season, probably being highest when the new leaves have fully expanded in the spring and increasingly lower as the leaves senesce toward winter. In contrast, *Larrea tridentata* maintains a relatively high potential photosynthetic activity year round. An indication of the potential dynamic change in photosynthetic characteristics of *Prosopis glandulosa* is the large change in leaf specific weight during the course of the year. New spring leaves have a specific weight of less than 4 mg per square centimeter. By fall, the leaves may attain a specific weight of nearly 17 mg per square centimeter which approximate the average specific weights of mediterranean evergreen sclerophyllous leaves. No such change is noted in the leaves of *Larrea*, which maintain a specific weight in Death Valley of about 12 mg per square centimeter for fully expanded leaves throughout the course of the year (Mooney, Björkman, Ehleringer, and Berry, unpublished data).

A portion of this change in leaf specific weight of *Prosopis glandulosa* leaves during the year is no doubt due to the progressive thickening of the cuticle with age. Interestingly, mature leaves of *Prosopis velutina*, which are deciduous like those of the honey mesquite, have a cuticle to cell wall ratio of six to one (6:1) whereas the leaves of the evergreen *Larrea* have a ratio inverse to this (Hull et al., 1975). It would thus appear that in Death Valley at least, new leaves emerge on *Prosopis* in the spring after the soil moisture has been recharged. Photosynthetic rates are high during this period. As temperatures increase, plants transport increasing amounts of water from the soil to the atmosphere *via* transpiration and consequently come under continually increasing water stress. The leaves apparently adjust not only physiologically to this stress by means of the stomatal behavior described earlier, but also morphologically by increasing their cuticular thickness as the season progresses. It is difficult to measure photosynthetic capacity in old leaves because of unusual stomatal sensitivity as well as the tendency of the leaves to fold when disturbed, but it can be assumed to decrease with advanced leaf age and water stress. Why *Prosopis* ultimately loses its leaves is not completely understood, but it is evident that their maintenance during the winter with very high water stress and cold temperatures is a liability which must exceed the cost of producing an entirely new canopy each spring.

ADAPTATIONS OF FRUITS, SEEDS, AND SEEDLINGS*

As with all plant species, the seed and seedling are the most vulnerable parts of the *Prosopis* life cycle. Because of the limited water supply of desert environments, it is particularly important that seeds be dispersed to the microsites in which they can successfully germinate and that seedlings possess adaptations for rapid and extensive development, particularly of root tissues. Our studies, combined with those of other researchers in both North and South America show that *Prosopis* species possess an interrelated suite of adaptations that ultimately promote seedling establishment.

Fruit Dispersal

All species of *Prosopis* section *Algarobia* (see Appendix for a description of the fruits of other sections) have fruits with three distinct layers (Figure

*O. T. Solbrig and B. B. Simpson.

6-2): a thin, leathery, variously colored exocarp, a semi-fleshy, often sweet mesocarp, and an inner, stony endocarp. Separated from one another inside the endocarp segments are several soft, oval, brown, seeds, each 5 to 10 mm long and 2 to 5 mm wide. The pods of the different species of this section are usually somewhat flattened and are straight or loosely coiled. In all cases, they fall from the tree as a single, indehiscent structure. Both the structure of the fruit and the high percentage of sugar and starch contained in the mesocarp indicate that the fruit is adapted for animal dispersal. Indeed, the sweet and slightly fleshy pods are known to be an attractive food source for animals, particularly mammals. Direct observations show that the fruits are avidly eaten by both wild and domesticated animals and examinations of scats of wild animals and feces of domesticated cows and goats reveal that many seeds, or even intact endocarps, pass through the digestive tract of these animals unharmed. Most of the data on the fate of consumed fruits comes from modern observations with cattle or other livestock which are today the principal consumers of *Prosopis* fruits. In feeding trials with horses, cows and ewes, 91 percent, 79 percent, and 16 percent respectively of the seeds consumed passed through the digestive tracts unharmed (Fisher et al., 1959). Germination trials with these seeds yielded 82 percent germination of those passed through a horse, 69 percent of those from cows and 25 percent of those from ewes (Haas et al., 1973). Shelled seeds which had not passed through any digestive tract had a germination percentage of only 26.

Originally, however, *Prosopis* fruits probably evolved with large native mammals of North and South America such as camelids, stegomastodons, notoungulates, and edentates, many of which went extinct at the end of the Pleistocene (see summaries in Martin, 1967; Patterson and Pascual, 1968; Pascual, 1970). Seeds in fossil dung show that mesquite was eaten in North America by the extinct Shasta ground sloth (Long et al., 1974) as well as by extinct sloths (as yet undetermined to genus) in Mendoza, Argentina (Martin, personal communication).

As described in Chapter 6, consumption by frugivores and passage through the digestive tract serves a dual function in that the seed is dispersed away from the parent tree and the internal seed parasites are killed by the digestive fluids. If the endocarp has been removed or cracked, and the seed not severely damaged (in fact, slight sacrification is beneficial) it is ready for germination. With sufficient moisture, seeds can germinate after only six hours at 34°C (Glendening and Paulsen, 1955; Lacher et al., 1963; Scifres and Brock, 1970; Eilberg, 1973; Solbrig and Cantino, 1975). The highest percentage germination recorded for *P. velutina* by Solbrig and Cantino (1975) was at temperatures near 30°C, but seeds will germinate at temperatures anywhere in the range of 20 to 40°C (Scifres and Brock, 1970; Eilberg, 1973). The effect of temperature is apparently to regulate the rate and extent of water uptake by germinating seeds (Scifres and Brock, 1971). At temperatures above or below the critical range, insufficient water is imbibed for germination. Although seeds will germinate on the surface of the soil, seedling survival depends on the seeds being covered by a thin layer of soil (about 1–2 cm). Since germination itself is not affected by light, the requirements of a soil cover appears to be related to selection for proper anchorage and maximum contact with humid soil (Scifres and Brock, 1970, 1972). Moisture is available in the desert at the surface layer of the soil only during a limited period of the year. An emergent seedling is completely dependent for a short period of time on the moisture surrounding the seed and natural selection has favored precise mechanisms that use a relatively narrow range of temperature and moisture as cues to insure germination only during the rainy season when conditions suitable for establishment are encountered. The high temperature requirement for germination is probably a result of the fact that *Prosopis* evolved in regions with summer rainfall and can be interpreted as a mechanism to prevent seeds from germinating during occasional off-season winter rains.

Seedling Establishment

Because of the short period during which the uppermost layers of the soil in desert environments are wet, plants, especially those dependent on relatively large supplies of moisture, must rapidly develop an efficient root system. Consequently, seedlings of these species exhibit rapid root elongation and high ratio of root to shoot growth relative to that of adult plants. When they first germinate, *Prosopis* seedlings have a prominent tap root and a pair of oval, somewhat fleshy cotyledons. No data are available for seedling development under completely natural field conditions, but simulated field studies have been made with seedlings of *P. velutina* at the Santa Rita Experimental Station in southeastern Arizona (Glendening and Paulsen, 1955). These studies indicate that root growth of seedlings in open areas protected from animal grazing was ten times greater than shoot growth (Table 2–1). Glendening and Paulsen concluded from their results that the slow shoot growth relative to root growth was due to an inhibition of stem elongation under conditions of moisture stress, rather than to a genetically determined tendency for precocious root growth. Their conclusion was based on the facts that, in the greenhouse with abundant soil moisture, there was less discrepancy between seedling root and shoot growth (Table 2–1) than in the cleared experimental plots, and that severe moisture stress is known to cause dormancy, near dormancy, or dieback of mesquite shoots (Glendening and Paulsen, 1955). It is difficult to see, however, why severe moisture stress would not have affected both root and shoot growth in the field. Moreover, although the ratios of root to shoot elongation were lower in the greenhouse than in the field, they remained relatively high. Our own observations (Solbrig, unpublished data) on several North and South American species and those of Haas et al. (1973) on *Prosopis glandulosa* under greenhouse conditions all showed very high rates of root growth. Nevertheless, the fact that root/shoot biomass ratios did change in the greenhouse indicates that more energy can be put into shoot development

TABLE 2-1 *Growth of seedlings of* Prosopis velutina *under Field and Greenhouse Conditions in SE Arizona*[1]

Time after Emergence of Cotyledons	Field			Greenhouse		
	Root Length (cm)[2]	Stem Length (cm)[2]	Ratio Root/Stem	Root Length (cm)[3]	Stem Length (cm)[3]	Ratio Root/Stem
10 days	7.87	2.03	3.87	–	–	–
15 days	11.43	3.30	3.46	–	–	–
25 days	33.27	6.61	5.03	–	–	–
30 days	–	–	–	–	6.30	–
60 days	38.1	6.30	6.40	–	–	–
5 months	–	–	–	68.32	60.45	1.13
9 months	51.05	5.33	9.58	–	–	–
12 months	–	–	–	121.92	17.14	7.11
38 months	84.83	11.40	7.4	–	–	–

1. Data from the Santa Rita Experimental Range. Adapted from Glendening and Paulsen, 1955.
2. Average values of forty-seven plants measured in each of three successive years.
3. Average values of twenty-two plants, except five month values for which only nine plants were used.

under favorable conditions and that there is, consequently, plasticity in seedling development that allows the maximum exploitation of available water
resources.

Despite adaptations for dispersal and germination, our observations at
both Andalgalá and Silver Bell suggest that seedling establishment is a rare
event under natural conditions. The only published data on seedling survival
are those of Haas et al. (1973) which report that there is a high seedling
mortality in populations of the honey mesquite in Texas. Fisher (personal
communication; Chapter 9) has observed that the Texas populations of
P. glandulosa often have distinct size (and presumably age) classes on a given
range site and postulated that seedling establishment occurs sporadically
during years of very favorable rainfall and only on sites where there is little
or no competitive vegetation cover (see also Scifres et al., 1971).

Field studies simulating natural conditions carried out with seeds and
seedlings of P. velutina showed that for this species as well, germination was
hampered and seedling mortality was high in patches with an established grass
cover (Glendening and Paulsen, 1955). In these studies, seeds of the velvet
mesquite were sown over a period of three years on patches from which the
vegetation had been removed and on patches with covers of various species
of grasses. Except for the exclusion of mammals, the experiments were undisturbed after sowing. Table 2-2 shows the differences in both germination
and survival of seedlings on grassy versus cleared plots. In all cases, germination and establishment was higher in areas free of competing grass species.
The germination and survival of the seeds and seedlings on the cleared patches
was probably higher in the experimental range than under natural conditions
at our site near Silver Bell because the Santa Rita Range has a higher annual
rainfall than the Silver Bell site and seedlings would be exposed to natural
mammalian herbivores. This study does, however, have important implica-
tions in the explanation of the spread of mesquite on to previously vege-
tated soils that have become denuded by overgrazing (Chapter 9).

TABLE 2-2 *Effects of Perennial Grass Cover on Germination and Survival
of* Prosopis velutina[1]

Site and Cover	Percent Germination[2]	Percent Survival to 1 year[2]
Bouteloua eriopoda (Black gamma)	2	7
Control (cleared)	50	71
Trichachne californica (Cotton grass)	7	18
Control (cleared)	56	80
Muhlenbergia porteri (Bush muhly)	0.1	0
Control (cleared)	43	66

1. Data from Santa Rita Experimental Range. Adapted from Glendening and Paulsen,
   1955.
2. Average values from a three-year study.

Once the seedling root reaches a permanent or semipermanent source of water, chances of survival are greatly increased if the seedling receives sufficient sunlight. Haas et al. (1973) report that newly germinated seedlings can not tolerate shade, and we found in both of our study areas that there were no seedlings under tree crowns. It is not known exactly how many years it takes from successful germination until the first production of flowers. Haas et al. (1973) reported about three years for the honey mesquite, but our field observations lead us to believe that in the desert under natural circumstances, reproductive maturity is reached much later.

SUMMARY

In many of their characteristics, individuals of *Prosopis* exhibit plasticity that is well adapted to desert environments with unpredictable moisture conditions. Because of the low total annual rainfall and the distribution of moisture in the soil, *Prosopis* behaves primarily as a phreatophyte in both of the study areas, but it has a root system capable of utilizing moisture in the upper layers of the soil. Likewise, its canopy and leaf morphology are similar to those of mesophytic plants but individuals show short-term physiological responses comparable to those of xerophytes. Consequently, *Prosopis* follows a pattern of efficient photosynthesis during the wetter/warm parts of the year, but usually loses its leaves during the dry/cold winter. In subtropical areas such as the Chaco and extreme southern Texas, *Prosopis ruscifolia* and *P. glandulosa* (respectively) often retain some leaves throughout the winter (Fisher, personal communication).

Patterns of seed dispersal and seedling establishment show adaptations for transport away from the parent tree and germination only after the reception of environmental cues correlated with favorable conditions for *Prosopis* growth. Seedlings are capable of rapidly producing a substantial root system, but growth patterns of seedlings, like those of the adult, appear to change in response to moisture conditions. In the next chapter we investigate components of variability not caused by individual plasticity. We examine first morphological and then genic and cytological variation present among individuals within populations and across geographic gradients.

PATTERNS OF VARIATION

O. T. Solbrig

K. Bawa

N. J. Carman

J. H. Hunziker

C. A. Naranjo

R. A. Palacios

L. Poggio

B. B. Simpson

Chapter 3

An important aspect of the success of any species is its ability to vary in response to both spatial and temporal fluctuations. Plants survive in fluctuating environments by possessing two fundamentally different, but not mutually exclusive, sets of attributes. First, individuals can respond phenotypically to a change in environmental conditions. This quality, known as phenotypic plasticity, is particularly effective if fluctuations occur at intervals on the order of, or shorter than, the life span of the individual. Another mechanism is the possession of a large supply of intrapopulational genetic variation that, by recombination, assures the production of variable offspring. For this latter mechanism to work, there must be more than one allele per gene and plants must outbreed to some extent (Darlington, 1939; Grant, 1958). Genetic variability is a more effective way than phenotypic plasticity for coping with environments that are patchy in time and space, and is a prerequisite for long-term evolution in a changing environment (Mather, 1943; Stebbins, 1950; Levins, 1968). Genetic variation can also be introduced into populations by occasional hybridization between species and subsequent backcrossing of the hybrids.

In this chapter we investigate some of the patterns of variation in species of *Prosopis* that could be associated with their success in different environments. We first examine the variability of some leaf and fruit characters and subsequently that of two classes of secondary plant chemicals. These studies estimate the degree of intraspecific and interspecific variability in the phenotype. However, variation in these characters could result from either phenotypic plasticity or from genetic divergence of individuals in different populations. After a discussion of these results, we then present data from studies on isozyme variation exhibited by different species, which permits us to make an estimate of the degree of genic variation in *Prosopis* taxa. Finally, we describe the cytology of most of the species of the genus and show the extent of interspecific hybridization and introgression between taxa. Throughout the discussions, we compare the species of *Prosopis* found at our two desert scrub study areas to one another, and to other members of the genus.

VARIATION IN GROSS MORPHOLOGY*

As pointed out in Chapters 1 and 2, numerous features of leaf morphology are associated with adaptations of plants to desert environments. In order to assess the range of potential phenotypic variability in these morphological characters, we analyzed populations of *Prosopis velutina* (plus *P. glandulosa*),

*B. B. Simpson.

FIGURE 3-1. *A lone tree of* Prosopis chilensis *stands among the rocks of the Rio Santa Maria near Cafayate, Salta, north of Andalgalá. Variation in growth form, leaf and fruit morphology combine with adaptations in physiology to promote the survival of* Prosopis *in desert environments.*

P. chilensis, and *P. flexuosa* across geographical gradients that extended be-yond the actual study sites. Since these geographical gradients are associated with environmental gradients, it was possible to infer if morphological varia-tion exhibited by the *Prosopis* species indicated similar responses to environ-mental factors in the two desert regions.

For this analysis, we sampled populations over large areas and looked for correlations between various characters and between different characters and physical parameters of the environment. We carried out most of the detailed studies on *Prosopis velutina* and adjacent populations of *P. glandulosa.* For these species, we took samples of three to five trees (two samples per tree) of *Prosopis* along a transect from Yuma, Arizona, to Fort Hancock, Texas (Fig-ure 3-2). Each sample from each tree included leaf material, fruits, and, when possible, flowers. After drying, we measured or scored the samples from each tree for (Figure 3-3):

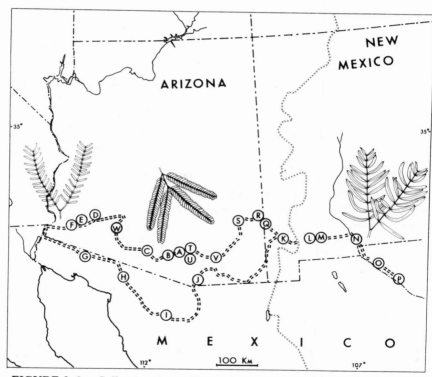

FIGURE 3-2. *Collecting sites of* Prosopis *across a geographical gradient from Yuma, Arizona, to Fort Hancock, Texas, U.S.A. Samples were taken from three to five trees from each locality during the late summer of 1973. The leaf on the far left is* P. glandulosa var. torreyana, *the one in the center,* P. velutina *and that on the right* P. glandulosa var. glandulosa. *The thin line of points indicates the continental divide.*

1. number of leaves per node
2. length of the leaf
3. length of the petiole
4. number of pairs of pinnae
5. number of pairs of leaflets
6. the length of the largest leaflet
7. width of the largest leaflet
8. width of the fruit (suture to suture)
9. length of the fruit
10. length of the largest pinna
11. width of the largest pinna
12. the distance between the fifth and sixth leaflet of a pinna
13. the thickest part of the fruit (with sutures medial to the measurement)
14. the ratio of the thinnest to the thickest part of the fruit
15. the ratio of the widest to the narrowest part of the fruit.

As can be seen in the Appendix, many of these characters are also used for distinguishing different species. The use of the ratios of the fruit measurements gave an estimate of the amount of beading (Figure 1–11) of the pod.

Once all of the characters were measured, correlations were sought between all pairs of characters and between different characters and longitude. Many of the characters were significantly and positively correlated ($p < 0.01$) with one another, indicating that they are inherited as character complexes. For example, the longest leaflets were also the widest, and the leaves with the greatest number of pinnae also had the largest number of leaflets. Some of the other correlations found were much less obvious in the field. The thickness and the width of the fruit were positively and significantly ($p < 0.01$) correlated with

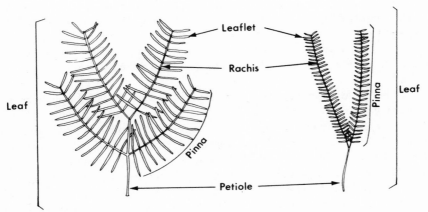

FIGURE 3–3. *Nomenclature of the parts of the leaves measured for the statistical analysis of morphological variation and for the species descriptions in the Key (see Appendix).*

the size of the leaves (and leaflets). In addition, the number of pairs of pinnae and leaflets were negatively correlated ($p < 0.05$) with the size of the leaf and the thickness of the fruit.

A clustering analysis using the fifteen characters recorded is shown in Figure 3-4. A gradual sequence of populations following the geographical gradient is evident (Figure 3-2), with a break at about the longitude of the continental divide. As shown in Figure 9-4, this area has a low density of *Prosopis* populations. The area is mountainous, supports a good grass cover, and appears to be relatively unsuitable for *Prosopis* growth. As seen in Figure 1-12, the same area constitutes the boundary between the ranges of *P. velutina* and *P. glandulosa* var. *glandulosa*.

Rainfall in the southwestern United States where the transect was made tends to increase as longitude decreases (west to east). Thus a high negative correlation of a character with longitude indicates a high negative correlation with precipitation, or a positive correlation with increasing aridity. Several of the original characters were significantly correlated with longitude ($p < 0.01$). These correlations showed that, going from east to west, populations of *Prosopis* have smaller leaves and leaflets, a larger number of leaflets, and thinner, narrower fruits. Figure 3-2 shows the extreme leaf types found at parts of the transect. Similar correlations of leaf characters with longitude across a different geographical area were noted by Graham (1960) and Johnston (1962), but fruit characters were not discussed.

When we examine these morphological changes in the light of increasing aridity from east to west, it appears that as the climate becomes drier, there is selection for the leaf surface areas to be broken down into a large number of small leaflets (see Chapter 2). Reduction in leaf size and the tendency to have dissected leaves has been shown to occur in numerous desert plant species (Gates, 1968). Small leaf size is advantageous in desert environments because it reduces the heat load of the leaves although at the same time it causes increased evaporation (Gates, 1968; Mooney, 1972; see also Chapter 2). The trend for the fruits to become thinner and more narrow as conditions become drier might reflect the fact that less energy is available to put into fruit production or that fruits are simply less hydrated where moisture is scarce.

In South America, samples from populations were collected over an area extending from the Province of Salta at about 26° S to the Province of La Rioja (29° S) and in an east-west direction from 66° W to 68° W. These populations included plants of *Prosopis chilensis, P. flexuosa, P. alba,* and *P. nigra.* Plants from the samples along this transect were not measured, but the samples were laid out and visually examined for trends in the change of morphological characters. In general, such things as leaf size, leaflet number, and fruit shape can readily be seen and changes are apparent when samples are viewed together. An actual statistical analysis would have been hard to perform because the terrain across which the Argentine samples were taken differs from that in the southwestern United States. Instead of a series of

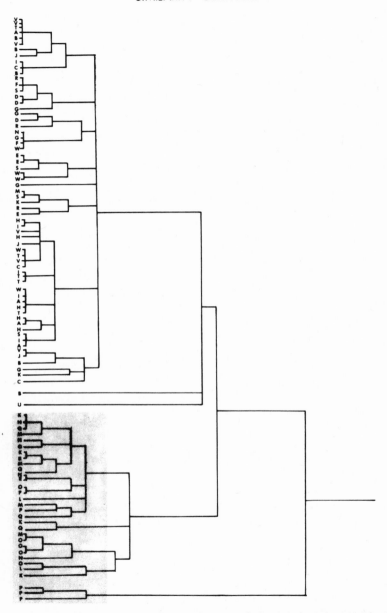

FIGURE 3-4. *A clustering of the trees sampled along the transect shown in Figure 3-2. The clustering is based on morphological similarity. The individuals in the cluster shaded in gray would be considered (for the most part) to belong to* P. glandulosa *var.* glandulosa *and the remaining populations to* P. velutina *(exceptions occur in both clusters). No pure population of* P. glandulosa *var.* torreyana *were sampled.*

undulating basins meandering around low mountains, which present a rather straightforward environmental gradient, the Argentine transect crossed large mountain chains unsuitable for *Prosopis* growth. The populations collected were thus rather isolated from one another and located in deep valleys. Nevertheless, on a very gross level, we know that north of Andalgalá *Prosopis* grows at elevations higher than in the Bolsón de Pipanaco and that south and west of the study area the climate becomes drier. The broad leaved *Prosopis chilensis* correspondingly becomes less and less common both to the north and to the south of Andalgalá and is replaced by species with smaller, more numerous leaflets (see Figure 1-14). Likewise, the species of *Prosopis* associated with these drier areas have thinner fruits than those of *P. chilensis.*

Patterns of morphological variation in terms of fruit and leaf characters seem to vary in similar ways in the *Prosopis* species of our study areas and in neighboring taxa. Large leaves with few leaflets and thick, comparatively fleshy fruits are associated in both regions with the more moist parts of a habitat gradient. These characters seem to be genetically based and are often associated in "character complexes."

VARIATION IN NATURAL PRODUCTS CHEMISTRY*

Our results of an analysis of two classes of secondary plant chemicals differ from those obtained from the study of morphological variation. The groups of chemicals analyzed were flavonoids and free amino acids. Both groups of chemicals commonly called "secondary plant products" are monomeric organic compounds with low molecular weights (more than 100 but less than 1000). These micromolecular components generally have restricted distributions among vascular and non-vascular land plants which makes them useful as characters in plant systematics. The extent to which natural products are synthesized and accumulated into cellular or extracellular pools is determined by the developmental stage of the life cycle, the organ or tissue, the growing season, and, naturally, the evolutionary history (genetic background) of the plant. On a dry weight basis, the amount of plant material incorporated in secondary plant compounds can vary from 1 to 30 percent.

For a long time, the great diversity of natural products was a puzzle for biochemists and physiologists since no function for them could be found. Recently, Whittaker and Feeny (1971) have advanced the hypothesis that these chemicals serve as defenses against plant herbivores (see also Chapter 4).

The two major types of natural products produced and accumulated in the leaves are flavonoids and free, non-protein amino acids (Carman, 1973). Flavonoids are a particular class of phenolic compounds that have been extensively used in studies dealing with population variation and systematics of angiosperms. These compounds are particularly suitable for analysis be-

*N. J. Carman.

cause they are stable and can be obtained from extractions of dried (i.e., herbarium) material and can be analyzed in microquantities. In addition, flavonoids are structurally complex, allowing opportunities for functional group heterogeneity. Free, or "unusual," amino acids are amino acids that are not incorporated into proteins. Like flavonoids, these compounds can easily be extracted and identified and have been useful as characters in previous subgeneric studies of other legume genera such as *Lathyrus* (Bell, 1962), *Vicia* (Bell and Tiramanna, 1965), *Astragalus* (Dunnill and Fowden, 1967) and *Acacia* (Senerviratne and Fowden, 1968).

Flavonoids of all the populations of *Prosopis velutina* and *P. glandulosa* sampled for morphological variation (Figure 3-2) as well as populations of numerous other species were extracted in alcohol from dried material. Extracts were placed on chromatography paper and run in two directions (Carman, 1973). The compounds represented by each spot that appeared on the chromatographs were eluted and identified using a combination of ultraviolet and mass spectroscopy. Eleven flavonoids were separated from crude extracts using sephadex and polyamide chromatography columns and identified by Nuclear Magnetic Resonance spectroscopy. Hydrolized extracts were used to determine the sugars of flavonoid glucosides.

The combined results showed that while species of *Prosopis* differed from one another in the number and kinds of flavonoids (Figure 3-5), and that the various populations within a species showed differences, there was no correlation of flavonoid variation with either the pattern of morphological variation described earlier in this chapter, or with any apparent geographical variable.

Amino acids were extracted from unground leaves of dried branches of the same Texas-Yuma transect populations analyzed for morphological variation and from numerous other North and South American populations. Samples from these populations were soaked for a week in methanol and water (Carman, 1973) and the filtered, concentrated extracts were used for two dimensional chromatography (Figure 3-6). All plant samples tested, regardless of species and/or population, were found to have high concentrations of the same two free amino acids: pipecolic acid and 4-hydroxy pipecolic acid. No quantitative differences in these two compounds were found in the various samples. In addition to these two free amino acids, proline, a protein amino acid, was found in large concentrations in all samples. However, using fresh material, it was found that pipecolic acid and proline are minor constituents of the total amino acid profiles (Cates and Rhoades, in press). Conceivably, the appearance of large quantities of these two amino acids in dried material is the result of a wilting response. Proline has been shown to increase in cells during water stress in *Solanum* (Levitt, 1972) and is postulated to be a "storage" compound for free NH_3 during times of water shortage. I have suggested (Carman, 1973) that pipecolic acids, since they are higher homologues of proline (although the product of an independent biosynthetic pathway), may serve a similar function. It is also possible that because of their analogous

Species	1 quercetin 3-OMe	2 apigenin 8-glucoside	3 apigenin 6-glucoside	4 quercetin 3-glucoside	5 quercetin 3-rhamnoside	6 myricetin 3-rhamnoside	7 quercetin 3-rutinoside	8 luteolin 7-glucoside	9 quercetin 3',3-diOMe	10 luteolin 3'-OMe	11 isorhamnetin 3-glucoside	12 apigenin 6,8-diglycoside	13 apigenin 6,8-diglycoside	14 myricetin 3-glucoside	15 myricetin 3',5'-diOMe, 3-rha	16 myricetin 3',5'-diOMe, 3-glu	17 isorhamnetin 3-rhamnoside ?	18 apigenin 6,8-glycoside	19 apigenin 6,8-glycoside	20 luteolin	21 kaempferol 3-OMe	22 quercetin
P. glandulosa									○	○	●	●	○									
P. velutina									○	○	●	●	○									
P. laevigata									○	○	●	●	○									
P. juliflora									○	○	●	●	○									
P. caldenia																						
P. flexuosa		○	○	●		○					○	○										
P. algarobia	○	○		●							●	●										
P. nigra	○	●	●	○		○					○											
P. alba	○	●	●	●	●	○	●													○	○	○
P. chilensis	○	●	●	●	●	○	●													○	○	○
P. alpataco						○					●	●	●									
P. ruscifolia											●	●	○									
P. sericantha											●	●	○									
P. kuntzei		●	●	●						○	○											
P. humilis		●	●	○	○						●	●						○				
P. ruizleali						●		○						○			○					
P. argentina		○	○		○	●								○								
P. reptans var. ciner.		●	●		●		●											○	○	○		
P. reptans var. reptans		●	●		●		●				○							○	○	○		
P. strombulifera		●	●		●		●											○	○	○		
P. torquata						●									●	●	●	○				
P. ferox								○														
P̄. cineraria								○														
P. farcata								○														

FIGURE 3-5. *Occurrence of various flavonoids in different species of* Prosopis. *The species are grouped by taxonomic affinities (see Appendix). Solid circles indicate compounds present in large amounts: open circles designated flavonoids in moderate or trace amounts.*

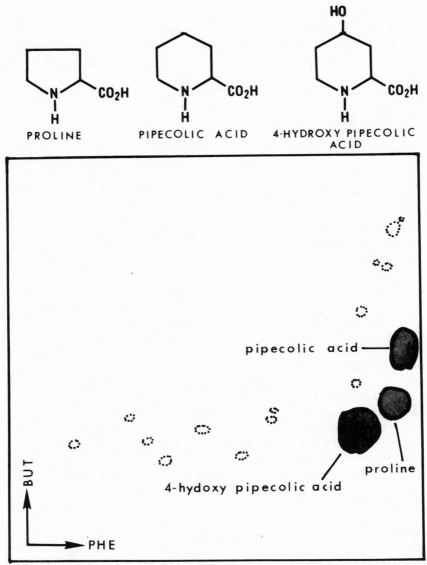

FIGURE 3-6. *The structure of the non-protein amino acids found in leaves of* Prosopis *and a diagrammatic representation of a chromatogram used for their identification.*

structure, pipecolic acids are incorporated in place of proline into the proteins of herbivorous animals and thus serve as chemical anti-herbivore defenses against grazing animals (see Chapter 4).

The study of natural products chemistry has shown that populations of *Prosopis* exhibit variation and that in some cases (e.g., the changing concentrations of pipecolic acid and proline) the variation reflects a plastic response to changing environmental conditions. However, the consistent patterns of flavonoid distributions in various species suggests that their production is genetically fixed within species and is not correlated with known local environmental factors.

GENIC VARIATION*

In order to make any statements about the genetic basis underlying the apparent phenotypic variation, or the potential effects of recombination, it is necessary to ascertain the degree of genic variability present in populations. Until recently, the elucidation of genetic variation involved crossing experiments which could, in general, be performed easily only on herbaceous plants with short generation times. The introduction of the isozyme technique (Hubby and Lewontin, 1966; Brewer, 1970) has provided an assay for genic variation that is easy to perform and does not require time consuming breeding trials.

The theory behind the use of this technique is as follows. Under the accepted hypothesis of gene function, every cistron codes for an enzyme (or polypeptide chain). The various alleles at a genetic locus (of a single gene) will all code for one particular enzyme such as alcohol dehydrogenase, but they can differ in one or more of their coding amino acid triplets. These codon differences produce polypeptide chains that differ in their chemical structure without differing in their primary function. Enzymes produced by alleles of the same locus and which consequently have similar functions but different structures are called isozymes, isoenzymes or allozymes (Markert, 1968). Using modern techniques, it is possible to separate and identify specific isozymes present in an extract of cellular material of a single individual.

Since each diploid organism has complementary sets of chromosomes, each individual has two alleles for each locus. If the organism is homozygous, the two alleles will produce the same isozyme. If the alleles differ, it is often possible to distinguish heterozygotes by extracting plant material for enzymes and testing their patterns. However, alleles can also produce isozymes that are different but appear similar because the enzymes lack sufficient physiochemical change to allow their detection. Nevertheless, by testing a large number of plants for a large number of enzymes, it is possible to arrive at an estimate of the number of alleles that constitute each locus and percentage of homozygous and heterozygous individuals in a population. We assume that the detection of a large amount of variation in the enzymes of the population samples indicates considerable total genetic variation in the entire population.

The technique we used to separate the isozymes is dependent on the surface charge of the molecules. Extracts of cellular proteins were embedded in

*O. T. Solbrig and K. Bawa.

the middle of a slab of starch gel that was placed in an electric field. The flowing electric current caused the isozymes to move along the gel toward the anode (positive pole) or cathode (negative pole) at speeds that varied according to the type and intensity of the charge of the protein and the size and shape of the molecule. After a designated time, the current was stopped and the slabs of gel were placed in a solution containing a soluble dye. Wherever there was an isozyme on the gel, the dye was oxidized (or reduced, according to the system used) and precipitated as visible bands. The patterns produced on the gel were scanned to determine whether there was one or more bands (i.e., isozymes) per individual and whether the plant was thus homozygous or heterozygous.

In this way we elucidated (Solbrig and Bawa, 1975) the isozyme variation for four enzyme systems of fifteen *Prosopis* species including *P. chilensis, P. flexuosa,* and *P. velutina.* Although the samples were too small to give precise determinations of gene frequencies, they did provide indirect measures that allowed estimation of levels of heterozygosity and the degree of genotypic polymorphism in the populations sampled.

Three measures were used for the calculation of the estimated variation: the isozyme diversity index, H' (the Shannon-Weaver measure of diversity or population evenness, see page 146); the ratio of H' divided by $H'max$ (the maximum value of H' for that number of samples); and a direct measure of the frequency of all possible combinations of isozyme bands (the "zymogram phenotype"). These measures taken together allowed an estimation of the number of genotypes in various populations, even though only four enzyme systems were analyzed. The number of loci involved, however, probably exceeded ten (Solbrig and Bawa, 1975).

Our results indicate that individuals of *Prosopis velutina, P. chilensis,* and *P. flexuosa* are heterozygous for part of their genomes. Extrapolation of the small number of loci sampled would indicate levels of genetic polymorphism as high as that observed to date in any angiosperm, and higher than those reported for *Avena* (Graminae) (Jain, 1969), *Leavenworthia* (Cruciferae) (Solbrig, 1972), or *Stephanomeria* (Compositae) (Gottlieb, 1973). Most of the other species of *Prosopis* analyzed were also polymorphic, but some exceptional populations of different taxa were found.

Three isolated populations of *Prosopis juliflora* from Colombia and Venezuela showed significantly lower levels of variation from populations in the main part of the range in western Central America. All three populations of *P. tamarugo* sampled (see Chapter 2 for a discussion of this unique species) also showed low levels of genic variation. Finally, many populations of *P. ruscifolia* in the Chaco of eastern Argentina and of *P. glandulosa* in western Texas, U.S.A., exhibited low levels of heterozygosity. Both of these last two species are "weedy" and have undergone recent population expansions (Chapters 9 and 10 and Figure 9-2 and 9-6; Morello et al., 1971; Fisher et al., 1973a).

The patterns of genetic variation detected through the analyses of isozymes indicates that populations of many species of *Prosopis,* including the mesquites at the two desert scrub study sites, are polymorphic. In contrast, some populations from extreme environments, narrowly isolated areas, or taxa that have recently undergone population explosions show comparatively low˙levels of genic variation.

CHROMOSOMAL CYTOLOGY AND HYBRIDIZATION*

In addition to genetic heterozygosity maintained by outcrossing within a population of a species, variation can be introduced into a population by hybridization with subsequent backcrossing. Another mechanism by which variation is ultimately introduced into a species is by mutation of alleles in the duplicate set of chromosomes following polyploidy. Earlier reports of polyploidy in species of *Prosopis* (Atchinson, 1951) and areas of hybridization (Stuckert, 1900; Burkart, 1937, 1940, 1943, 1952; Graham, 1960; Johnston, 1962; Morello et al., 1971) encouraged us to carry out a study of the cytology of most of the species of *Prosopis* and many apparent cases of hybridization (Hunziker et al., 1975).

As a result of these studies, plus previous reports, we now know the chromosome numbers of twenty-eight of the forty-four species of *Prosopis* including those in both of the desert scrub study areas (Table 3-1). All taxa studied thus far are diploid with a haploid number of fourteen ($2n=28$) except for *P. juliflora* which appears to have both diploid and tetraploid races (Table 3-1). This is a very low frequency of polyploidy, when compared to the value of 20% obtained when all of the species of the Leguminosae known cytologically are tabulated (Bir and Sidhu, 1967). We attribute the scarcity of polyploidy in *Prosopis* to the fact that most species do not possess life history and genetic characteristics that are associated with polyploidy (Grant, 1971). These features are: long life combined with vegetative reproduction, a pattern of primary speciation followed by chromosome repatterning, and a high frequency of interspecific hybridization. While species of *Prosopis* are long lived, there is little or no vegetative reproduction in most of the species (some of the small members of section *Strombocarpa* are exceptions). Moreover, as illustrated below, there appears to be extremely little chromosome repatterning accompanying speciation in the genus, a phenomenon that probably negated any effect that hybridization might have otherwise had on polyploidy. Interspecific hybridization among *Prosopis* species does appear to be fairly common, but the lack of sterility barriers seems to have resulted in backcrossing rather than the production of polyploids.

Chromosome counts of most of the species were obtained from mitotic squashes of root tips. In many cases, a high percentage of polysomatic cells were observed with double or quadruple the number of chromosomes. How-

*J. H. Hunziker, C. A. Naranjo, R. A. Palacios, and L. Poggio.

TABLE 3-1 *Chromosome Numbers of* Prosopis *Species and Putative Hybrids*[†]

Taxon or Hybrid (see Appendix for authors)	Meiotic Number n	Meiotic Number 2n	Origin of Plant Material
Section *Algarobia*			
P. alba	14[c]		Argentina, Formosa
		28[b,c]	Argentina, Córdoba
P. affinis	14[c]	28 (56)[c]	Argentina, Entre Ríos
(as P. algarrobilla)	14[c]		Argentina, La Pampa
P. alpataco	14[c]	28 (56)[b,c]	Argentina, Mendoza
P. caldenia		28 (56)[a]	Argentina, Mendoza
P. campestris		28 (56)[b]	Argentina, Córdoba
*P. chilensis		28 (56)[a]	Argentina, Mendoza
*P. flexuosa	14[c]		Argentina, La Pampa
		28 (56)[a]	Argentina, Mendoza
P. glandulosa	14[d]		Pakistan, Indus Delta (cultivated)
var. glandulosa		28 (56, ca 112)[a]	U.S.A., Texas
var. torreyana		28 (56)[a]	U.S.A., California (cultivated)
P. hassleri		28 (56)[c]	Argentina, Formosa
P. humilis		28 (56, ca 112)[a]	Argentina, Córdoba
P. juliflora	14[c]	28 (56, ca 112)[c]	Colombia, Bolivar
		28 (56, ca 112)[c]	Brazil, Río Grande de Norte
		(ca 112)[c]	Colombia, Tolima
		56 (ca 112)[c]	Haiti, L'Ouste
		56 (ca 112)[c]	Venezuela, Lara
P. kuntzei		28 (56, ca 112)[a]	Argentina, Santiago del Estero
P. laevigata		28[c,e,g,h]	Mexico
P. nigra	14[c]		Argentina, Formosa
	14[c]	28 (56, ca 112)[c]	Argentina, Entre Ríos
P. patagonica		28 (56)[c]	Argentina, Río Negro
P. ruiz-leali	14[c]		Argentina, Catamarca
P. ruscifolia	14[c]		Argentina, Formosa
		28 (56)[f]	Argentina, Santiago del Estero
P. sericantha		28 (56, ca 112)[a]	Argentina, Mendoza
*P. velutina		28 (56)[a]	U.S.A., California (cultivated)
Section *Monilicarpa*			
P. argentina		28[b]	Argentina, Mendoza
Section *Strombocarpa*			
P. pubescens		28 (56)[a]	U.S.A., California
P. aff. reptans		18 (56, ca 112)[a]	Argentina, Mendoza
P. strombulifera		28[i]	Argentina, Mendoza
P. tamarugo		28[c]	Chile, Tarapacá
P. torquata		28 (56)[a]	Argentina, La Rioja
P. ferox		28 (56, ca 112)[a,b]	Argentina, Jujuy
Section *Adenopsis*			
P. farcta		28 (56)[a]	Iran, Tehran
Hybrids			
P. alba × P. nigra ?	14[c]		Argentina, Tucuman
P. alba × P. flexuosa ?		28 (56)[c]	Argentina, Catamarca
P. hassleri × P. ruscifolia ?	14[c]		Argentina, Formosa
P. vinalillo			Argentina, Formosa
(P. ruscifolia × P. alba ?)			
	14[c]	28 (56)[c]	

[†]Numbers in parentheses indicate polyploid cells due to polysomaty.
*Species of *Prosopis* found at the study sites.
b. Covas and Schnack, 1946, 1947.
c. Hunziker et al., 1975.
d. Baquar et al., 1966.

e. Johnston, 1962.
f. Covas, 1950.
g. Johnston, 1962.
h. Isley, 1972.
i. Schnack and Covas, 1947.

57

ever, meiosis was also observed in several species and found in all cases to be regular with fourteen bivalents at metaphase 1. The observations of the meiosis of putative hybrids between *P. alba* X *P. nigra* (?), *P. hassleri* with *P. ruscifolia* (?), and *P. ruscifolia* with *P. alba* (=*P. vinalillo* ?) showed regular meiosis with fourteen chromosome pairs. Karyotype morphology of many of the species and the putative hybrids were similar. The twenty-eight chromosome complement is composed of small, only slightly differentiated chromosomes. The chromosomes have median to subterminal centromeres, one pair of which displayed a terminal microsatellite in most species. In the case of the *P. ruscifolia* X *P. alba* hybrid, no significant differences were found between parental species and their apparent hybrids in regard to the frequency of numbers of paired bivalents, chiasmata per cell, or per chromosome pair. Hybrids also appeared to be completely or nearly fertile and exhibited morphological characters that were intermediate between those of the parental types.

In terms of chromosome number and morphology, therefore, there appear to be few genetic or chromosomal barriers to hybridization between various species of *Prosopis*. In addition to these cytological features, sympatry, partial overlap of flowering times, and little specific discrimination by pollinating insects (for *Prosopis* species within a section, see Chapter 5) would tend to facilitate hybridization. Although isolating mechanisms appear to be weak, it is still possible that other barriers such as degeneration of the hybrid progeny may prevent swamping of poorly isolated species by hybridization.

In addition to the cases we have investigated cytologically, we have observed many populations which contain apparent hybrids between species of *Prosopis* section *Algarobia*. In South America, this section has several species groups, one of which consists of sub-aphyallous trees with multinodal axillary or terminal spines. The two species of this group (Appendix) are not known to hybridize with any other members of the genus, but one intermediate specimen suggests that they might occasionally hybridize between themselves. Both of these taxa flower at the beginning of December while the remaining species of section *Algarobia* with which they are sympatric flower at the beginning of September.

The group of species of section *Algarobia* encompassing the species we have studied in detail have paired spines (occasionally appearing spineless) or have uninodal or solitary spines. We have identified at least eight different putative hybrids between the species of this group (Palacios and Naranjo, unpub.). These hybrids, all with high fertility and intermediate morphological features are: *P. affinis* X *P. alba* (?), *P. nigra* X *P. alba* (?), *P. flexuosa* X *P. nigra* (?), *P. hassleri* X *P. alba* (?), *P. ruscifolia* X *P. hassleri* (?), *P. ruscifolia* X *P. fiebrigii* (?), and *P. ruscifolia* X *P. alba* (?). Other South American reports of hybridization include mention of individuals with morphologies intermediate between *P. ruscifolia* and *P. hassleri* (Morello et al., 1971) in Formosa Province, Argentina; between *P. affinis* and *P. ruscifolia* in Paraguay (Fisher, personal communication); and between *P. caldenia* and *P. flexuosa* in La Pampa, Argentina (Covas, personal communication).

It is also possible that some of the rare species might have arisen by hybridization between relatively unrelated species within a section. Such a case might be *Prosopis abbreviata* (section *Strombocarpa*) which looks like an intermediate between *P. torquata* and *P. strombulifera* (or *P. reptans*). It has also been suggested that *P. ruscifolia* var. *parvifolia* (section *Algarobia*) is a hybrid derived from a cross between *P. ruscifolia* and *P. vinalillo* (Burkart, 1940), a species itself suspected of being of hybrid origin (Stuckert, 1900; Burkart, 1940). Finally, *Prosopis burkartii* (section *Strombocarpa*), a taxon recently described (Munoz P., 1971) from Tarapaca, Chile, may have resulted from a cross between *P. tamarugo* and *P. strombulifera*. Despite the apparent abundance of intrasectional hybridization, there are to date no examples of hybrids from crosses between species belonging to different sections in either North or South America.

Evidence for hybridization between the North American species of *Prosopis* belonging to section *Algarobia,* has been presented by several authors. Graham (1960) made a detailed study of morphological variation in mesquite from seventeen sites in northeastern Mexico and found intermediate individuals in a narrow zone that indicated the occurrence of hybridization between the long-leaved honey mesquite, *P. glandulosa* and the short-leaved, *P. laevigata.* Johnston (1962) later studied all of the North American species of section *Algarobia* and found what he called "contamination of populations" of *P. glandulosa* by "genes" of *P. velutina*. According to the author, some of the most confusing North American plants from a taxonomic point of view are members of this complex and are found in southern Sonora, Baja California, and northern Sinaloa. Plants in the vicinity of Guaymas, Sonora, and La Paz, Baja California, exhibited a combination of characters found in *P. articulata, P. glandulosa, P. velutina,* and, perhaps, even *P. juliflora.* Johnston pointed out that there is no cytological evidence of cross incompatibility and emphasized that the increase of mesquite in recent times (see Chapter 9) may have been a factor contributing to the blurring of formerly more natural discontinuous distributions by providing increased opportunities for hybridization. It is also possible that introgression of genes from *Prosopis laevigata* into the large leaved *P. glandulosa* var. *glandulosa* (Figure 3-2) is responsible for the tendency of many mesquites in the area of Big Bend, Texas, southward to central Chihuahua, Mexico, to have numerous small leaflets close together on the rachis.

SUMMARY

In these studies of *Prosopis,* we have investigated the variability of leaf and fruit characters, phenolics and free amino acids, isozymes and cytological features. We found that populations of many species of *Prosopis* are genetically polymorphic and exhibit variation in the phenotypic characters we investigated. The patterns of variation in the South American *Prosopis chilensis* and *P. flexuosa* were similar to those of the North American populations of *P. velutina* and *P. glandulosa*.

Leaf and fruit characters show clinal variation within species that can be correlated with conditions of aridity. Leaflets are larger and fruits are fleshier in mesic areas than in xeric ones. Phenolic compounds (flavonoids) do not show the same type of clinal variation, but each species appears to have a distinctive, species-specific array of chemical compounds.

All taxa studied thus far are diploid with a haploid number of 14 ($2n=28$) except for *P. juliflora* which appears to have both disploid and tetrapoloid races.

The lack of chromosomal differences and the absence of incompatibility barriers appear to account, in part, for the high incidence of interspecific hybridization among related species of the genus and doubtlessly has contributed to the variability found in *Prosopis* populations. However, the variability also appears to be the result of selection for genetic polymorphisms in populations that is advantageous for these long-lived perennials that inhabit a broad spectrum of relatively patchy micro and macrohabitats.

Not only the physical parameters of the environment, however, act as selective agents on species of *Prosopis*. Intensive studies of the leaf, flower, and fruit herbivores at the two study sites discussed in the next three chapters indicate that biotic interactions must also play a significant role in shaping many of the adaptive features we now see in the mesquite and algarrobos of Silver Bell and Andalgalá.

FIGURE 4-1. *Leaves of* Prosopis flexuosa *near Andalgalá, Argentina, preferred by two species of leaf herbivores,* Oiketicus geyeri (*Psychidae*) *and* Melipotis bisinuata (*Noctuidae*).

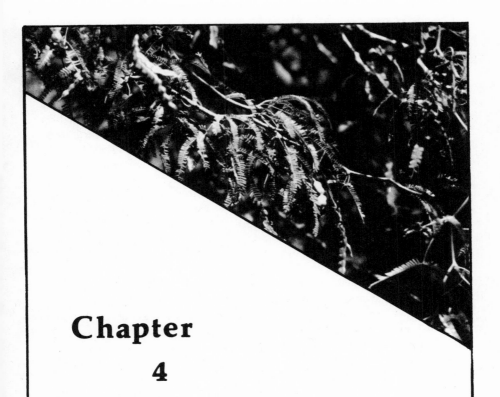

Chapter
4

PROSOPIS LEAVES AS
A RESOURCE FOR
INSECTS

R.G. Cates
D.F. Rhoades

The consumption of tissues, particularly leafy tissues, by primary herbivores is a fundamental process in all terrestrial ecosystems. Although often overlooked, insects are the most important plant herbivores on a global scale in terms of numbers of species, numbers of individuals, and amounts of plant tissue consumed. Plant taxa differ in the number of insects which feed on their leaves and the extent to which they are affected by invertebrate grazers. Likewise, insects differ in the number of plant taxa on which they feed. At the two desert sites, we found that species of *Prosopis* also varied in their numbers of leaf herbivores. Moreover the major insect herbivores differed in their specificity toward *Prosopis* and even in the types of leaf tissue which they preferentially consumed. We therefore undertook to answer the question, what are the relationships between the degree of specificity of a leaf herbivore, the kind of tissue it prefers, and the chemical and physical properties of its preferred tissue? Conversely, by what mechanisms does *Prosopis* limit or avoid predation?

It has been known for a long time that vertebrates, particularly domesticated species, avoid plants with mechanically irritating tissues and/or those with a "bad taste" or poisonous chemicals. Less is known about the ways in which plants avoid predation by invertebrates even though they are important consumers of plant vegetative tissues. Several studies strongly suggest that many of the secondary plant chemicals provide significant protection against such herbivores (Feeny, 1970; Harborne, 1972; Jones, 1972). It now also appears that the nature and extent to which plants use these deterrents may depend in part on their growth form and successional status (Cates and Orians, 1975). Few studies exist, however, pertaining to the distribution of secondary chemicals in various tissues of plants in relation to the feeding patterns of insect herbivores. We have elaborated elsewhere a general theory of plant-herbivore interactions (Rhoades and Cates, 1976; Cates et al., 1977) and will therefore limit ourselves here to a discussion of *Prosopis* and some of its major insect herbivores at the two study sites.

In our work with *Prosopis* and these herbivores, we have determined the degree of specialization of several species of insects on the three species of *Prosopis* near Andalgalá and of *P. velutina* at Silver Bell. Herbivores that consume only one species are considered to be specialists or monophagous, those that utilize a limited range of plant species are considered oligophagous, and insects feeding on a wide spectrum of plant taxa are designated as generalists or polyphagous. We then analyzed in detail the *Prosopis* tissue preferences of these herbivores and the correlations, if any, with plant chemistry.

DEGREE OF TAXONOMIC SPECIALIZATION
OF *PROSOPIS* HERBIVORES

The degree of dietary specialization among several insect species associated with *Prosopis* was studied intensively in the Bolsón de Pipanaco (km 1512, Figure 1-5) on a plot 5 km long and 75 m wide. The plot contained three

TABLE 4-1 *Field Observations of Leaf Herbivory by Insects Utilizing* Prosopis *Species at the Bolsón and Silver Bell Study Sites*

Insect species	Plant species	Plant density[1]	No. obser. per plant species	Percent of total observations
Semiothisa sp.	*Acacia aroma*[2]	11.1	650	28
(Geometridae)	*Cercidium praecox*	5.6	324	14
	A. furcatispina	3.0	580	25
	Prosopis chilensis	1.4	61	3
	P. torquata	1.3	653	28
	P. flexuosa	0.3	23	1
Oiketicus geyeri Berg	*Acacia aroma*	11.1	1049	83
(Psychidae)	*Cercidium praecox*	5.6	33	3
	Prosopis chilensis	1.4	53	4
	P. flexuosa	0.3	75	6
	Mimosa farinosa	0.3	52	4
Melipotis bisinuata	*Acacia aroma*	11.1	359	54
Felder & Rogenhoefer	*Prosopis chilensis*	1.4	61	9
(Noctuidae)	*P. flexuosa*	0.3	242	37
Undetermined Noctuid A (Noctuidae)	*Prosopis chilensis*	1.4	29	100
Brachyphatnus sp. (Pergidae)	*P. chilensis*	1.4	83	100
Epicauta arizonica Werner (Meloidae)	*P. velutina*	no data	216	100

1. Number of individuals per plot from 24 plots, each 50 X 50m on the study site.
2. All species are members of the Leguminosae. Authors are given in the Index to Scientific Names.

species of *Prosopis; P. chilensis, P. flexuosa,* and the screwbean *P. torquata,* as well as several other, primarily legume, species. Trees in the plot were sampled for insect herbivores (Table 4-1) twice weekly during the southern summers (January to March) of 1973 and 1974. Searching at night, we recorded the plant species being fed upon and the kind of leaf tissue preferred on each plant species. All the observations and experimentation on *Prosopis* leaf herbivores in the Bolsón study areas were concentrated on four lepidopteran larvae (*Semiothisa* sp., *Oiketicus geyeri, Melipotis bisinuata,* and an unidentified noctuid, species A) and the larvae of *Brachyphatnus* sp., a sawfly. The noctuid could not be determined to genus or species from the larvae and no adults were successfully reared. The species of sawfly also remains undetermined. At Silver Bell, less intensive studies were made. These centered primarily on adults of the meloid beetle, *Epicauta arizonica,* the major leaf chewing insect on *Prosopis velutina* during August 1973. This species was not found on *Prosopis* during the last part of July nor during the first two weeks of September. Tables 4-1 and 4-2 summarize these observations.

Table 4-1 includes not only a summary of the observations for each of the herbivore species on *Prosopis* and other hosts, but also the density of each plant species on the Argentine study plot. As can be seen two of the lepidopteran species, *Semiothisa* and *Oiketicus geyeri* proved to be generalized in their feeding patterns and fed on six and five plant species respectively. In contrast, *Melipotis bisinuata* was restricted to three host plant species. The remaining two species, Noctuid A and *Brachyphatnus*, were apparently specialists on *Prosopis chilensis* (Table 4-1), although in view of the low number of observations they may prove later to have a more broadly based diet. In our two year study, 1,168 individuals of the various legume tree and shrub host species were sampled on and off the study plots and Noctuid A and *Brachyphatnus* were always found with, and only found with, *Prosopis chilensis.* All of the herbivores were consistently present on *Prosopis* at the Bolsón site when leaves were present, but their abundance varied widely from year to year.

In discussing the feeding patterns of the *Prosopis* invertebrate herbivores, we will use primarily the relative preferences of the insects for various food plants. Preference ranking (Table 4-2) was determined by dividing the number of observations of each herbivore on a particular plant species eaten by that host plant's density and then normalizing over all plant species eaten by the herbivore. For convenience, the most highly preferred species was given a value of 100. This type of tabulation reflects the preference of the herbivores in relation to the availability of the food resource and removes bias introduced by differing plant species abundances.

A comparison of the percentage of total grazing observations for herbivores on individual plants (Table 4-1) with host plant preference ranking (Table 4-2) illustrates the advantages of this method of plant preference

TABLE 4-2 *Preference Ranking of Host Plants to* Prosopis *Herbivores at the Bolsón Study Site*

| Plant species | Herbivore species | | | | |
	Semiothisa sp.	*Oiketicus geyeri*	*Melipotis bisinuata*	Noctuid A[1]	*Brachyphatnus* sp.
Prosopis chilensis	9	15	5	100	100
P. torquata	100	0	0	0	0
P. flexuosa	0	100	100	0	0
Acacia aroma	12	38	4	0	0
A. furcatispina	38	0	0	0	0
Cercidium praecox	12	2	0	0	0
Mimosa farinosa	0	17	0	0	0

1. This lepidopteran species was not identified because no adults could be reared. Special thanks are extended to Drs. D. W. Duckworth, W. D. Field, R. W. Hodges, E. L. Todd, D. C. Ferguson, and D. M. Weisman for identification of the herbivores.

ranking. For example, the highest percentages of grazing observations for *Oiketicus geyeri* (86 percent), *Melipotis bisinuata* (54 percent), and *Semiothisa* (28 percent) are all on *Acacia aroma*. However, the preference rankings on *A. aroma* of *O. geyeri* (38), *M. bisinuata* (4) and *Semiothisa* (12) are all relatively low compared to the preferred host plant. The reason, of course, is that *A. aroma* is the dominant host plant species at our study site near Andalgalá. Thus, true host plant preference is not revealed by grazing observations until the differences in plant densities are taken into account as outlined above.

If we look first at the feeding preferences of *Semiothisa* (Table 4-2), we find that it prefers *Prosopis torquata* even though this species is relatively rare in the study area. The second most preferred species is *Acacia furcatispina*, a tree of intermediate abundance at the site, followed by *A. aroma, Cercidium praecox, Prosopis chilensis,* and *P. flexuosa*. The psychid *Oiketicus geyeri* preferred a different species of *Prosopis, P. flexuosa,* followed by *Acacia aroma, Mimosa farinosa, P. chilensis,* and finally, *Cercidium praecox*. In contrast to *Semiothisa, Oiketicus geyeri* does not eat *Prosopis torquata* at all. A third species, *Melipotis bisinuata,* also prefers *Prosopis flexuosa* but has a diet much more restricted than that of either *Semiothisa* or *O. geyeri*. Finally, the two remaining species, Noctuid A and *Brachyphatnus* sp. were both monophagous, grazing on only *P. chilensis*.

Phenological observations (LeClaire et al., 1973a,b; LeClaire and Brown, 1974a,b) indicate that the timing of leaf production is of little importance in determining herbivore preferences at the site near Andalgalá since leaf initiation is similar for both *Prosopis chilensis* and *P. flexuosa* in the Bolsón de Pipanaco (Solbrig and Cantino, 1975). The other plant species grazed (Table 4-1) are more dependent on the timing of summer rains and thus are more variable in the timing of leaf initiation. For example, few of the plants of *Acacia aroma, A. furcatispina, P. torquata,* and *Cercidium praecox* in the flats around the study site were in full leaf in the dry spring of November and December 1972. However, once the rains began, they leafed fully and quickly, suggesting that on the flats, legumes tend to leaf synchronously and consequently time of leaf emergence does not contribute to herbivore preferences between host plants.

Summarizing the data of breadth of diet and plant species preferences, we find that *Semiothisa* and *Oiketicus geyeri* have larvae that feed on six and five legume plant species respectively. The first, however, prefers the screwbean, *P. torquata,* and the latter, *P. flexuosa: Melipotis bisinuata* larvae also feed on both, but prefer *P. flexuosa*. Noctuid A and the sawfly are apparent specialists on *Prosopis chilensis*. In Arizona, plant densities were not determined so plant preferences could not be ranked in the same manner. However, as seen in Table 4-1, *Epicauta arizonica,* the species with which we worked, is monophagous on *P. velutina*. This preference was determined by presenting sprays from all of the woody legumes on the study plots to the beetles. At the time of our trials, only leaves of *Prosopis velutina* were eaten. We do not know if *E. arizonica* feeds on other plant species at different times of the year.

FEEDING PATTERNS OF HERBIVORES IN
RELATION TO LEAF AGE CLASSES

In addition to examining the plant hosts, we examined the preferences of the herbivores on *Prosopis* and associated legumes for various age classes of leaf tissue. These determinations were made in the field and tested in the laboratory. Leaf tissue types as we recognized them are shown in Figure 4–2. Leaves which were just beginning to grow with little apparent differentiation were classified as young leaves, and those which were well formed and completely differentiated but not yet hardened, as intermediate leaves. We termed mature leaves those that were well developed and tough but not senescent. We first estimated the amount of each of these tissue age classes of various plant species available to invertebrate herbivores and then defined the field preference of an herbivore as the ratio:

$$\frac{\text{number of insect feeding observations per tissue type}}{\text{proportion of tissue type available.}}$$

We determined the availability of the three age classes of leaf tissue in the field in the following manner. Ten randomly chosen sprays from ten different

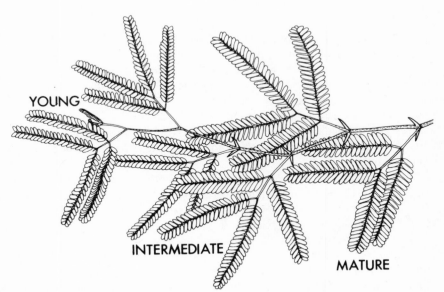

FIGURE 4–2. *A diagram of the leaf age classes used in determining herbivore grazing preferences. Young leaves were those just beginning to grow and with little or no differentiation. Intermediate leaves were well formed, but not yet hardened. Mature leaves were both well developed and tough, but not yet senescent.*

individuals of each host plant were clipped and the leaves from each separated into age classes. We then weighed each age class and calculated the percentage dry weight of each tissue type for all of the 100 sprays. Preference would be equal if age classes were eaten in proportion to our estimation of their availability. Relative field tissue preference (RTP) of each herbivore was calculated by assigning a value of 100 to the most highly preferred tissue and normalizing the other values.

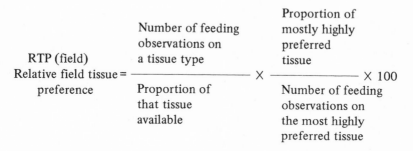

$$\text{RTP (field)} \atop \text{Relative field tissue} = \frac{\text{Number of feeding observations on a tissue type}}{\text{Proportion of that tissue available}} \times \frac{\text{Proportion of mostly highly preferred tissue}}{\text{Number of feeding observations on the most highly preferred tissue}} \times 100$$

Under laboratory conditions, preference was calculated as a proportion of the amount eaten relative to the known amount offered. In this case, the

$$\text{RTP (lab)} \atop \text{Relative laboratory} = \frac{\text{Amount of tissue eaten}}{\text{Amount of tissue offered}} \times \frac{\text{Amount of most preferred tissue offered}}{\text{Amount of most highly preferred tissue eaten}} \times 100$$

Even though there is considerable overlap in the plant species grazed by the herbivores studied in the Bolsón, there are distinct differences in their leaf tissue preferences (Table 4-3). Throughout this study, a significant difference is considered one for which $p \leq 0.05$. The statistic used was the Student "t" test unless specified otherwise.

The two herbivores with the broadest diets, *Semiothisa* and *Oiketicus geyeri* (six and five plant species respectively), prefer significantly ($p < 0.05$) mature leaf tissues of all of their hosts over either intermediate or young tissues. However, *Melipotis bisinuata* significantly prefers the young leaves of its three host species. Noctuid A, *Brachyphatnus* and *Epicauta arizonica* in Arizona, all of which appear to be specialized upon *Prosopis,* strongly and significantly prefer the youngest leaf tissues in the field and feed only slightly on intermediate and mature tissues in the laboratory (Table 4-3).

These results show that for the insects studied, polyphagous species prefer the mature leaves of their hosts whereas the oligophagous and monophagous species prefer the youngest leaves.

TABLE 4-3 *Leaf Tissue Preference of Herbivores on Their Host Plants from Field and Laboratory Experiments*[1]

	Leaf age class														
	P. flexuosa			*P. chilensis*			*P. torquata*			*A. aroma*			*A. furcatispina*		
	Y	I	M	Y	I	M	Y	I	M	Y	I	M	Y	I	M
Semiothisa sp.															
Field:															
No. feeding observations	0	—	23	11	51	516	71	—	2067	0	168	1379	20	—	1202
No. feeding observations	0	0	0.26	2.2	2.8	6.7	11.8	—	22.0	0	7.30	18.9	1.7	—	13.7
% tissue available	0	—	100	45	57	100	54	—	100	0	39	100	9	—	100
Relative tissue preference	—	—	—	33	47	100	—	—	—	—	—	—	—	—	—
Laboratory:															
Relative tissue preference	—	—	100	—	—	—	—	—	—	—	—	—	—	—	—
Oiketicus geyeri															
Field:															
No. feeding observations	0	1	72	0	4	41	—	—	—	9	74	658	—	—	—
No. feeding observations	0	0.03	0.8	0	0.22	0.53	—	—	—	2.2	3.2	9.4	—	—	—
% tissue available	0	4	100	0	42	100	—	—	—	24	36	100	—	—	—
Relative tissue preference	—	—	—	—	—	—	—	—	—	—	—	—	—	—	—
Melipotis bisinuata															
Field:															
No. feeding observations	212	2	6	108	4	0	—	—	—	193	52	6	—	—	—
No. feeding observations	30.3	0.7	0.7	22.0	0.22	0	—	—	—	27.6	2.3	0.08	—	—	—
% tissue available	100	2	2	100	2	0	—	—	—	100	7	0	—	—	—
Relative tissue preference	—	—	—	100	40	24	—	—	—	—	—	—	—	—	—
Laboratory:															
Relative tissue preference	100	—	—	—	—	—	—	—	—	—	—	—	—	—	—

TABLE 4-3 (Continued)

	P. flexuosa			P. chilensis			P. torquata			A. aroma			A. furcatispina		
	Y	I	M	Y	I	M	Y	I	M	Y	I	M	Y	I	M
Noctuid A.															
Field:															
No. feeding observations	—	—	—	29	0	0	—	—	—	—	—	—	—	—	—
No. feeding observations	—	—	—	5.8	0	0	—	—	—	—	—	—	—	—	—
% tissue available	—	—	—	100	0	0	—	—	—	—	—	—	—	—	—
Relative tissue preference	—	—	—				—	—	—	—	—	—	—	—	—
Laboratory:															
Relative tissue preference	—	—	—	100	10	6	—	—	—	—	—	—	—	—	—
Brachyphatnus sp.															
Field:															
No. feeding observations	—	—	—	83	0	0	—	—	—	—	—	—	—	—	—
No. feeding observations	—	—	—	16.6	0	0	—	—	—	—	—	—	—	—	—
% tissue available	—	—	—	100	0	0	—	—	—	—	—	—	—	—	—
Relative tissue preference	—	—	—				—	—	—	—	—	—	—	—	—
Laboratory:															
Relative tissue preference	—	—	—	100	0	0	—	—	—	—	—	—	—	—	—
Epicauta arizonica *(P. velutina)*															
Field:															
No. feeding observations	101	0	0												
No. feeding observations	20.2	0	0												
% tissue available	100	0	0												
Relative tissue preference															
Laboratory:															
Relative tissue preference	100	32	0												

1. Y = young; I = Intermediate; M = Mature leaf tissues   ☐ = Preferred tissue
Data from trees both on and off the study site.

69

NITROGEN ACCUMULATION AND MASS UTILIZATION

Feeding patterns were further investigated by determining the ability of the various herbivores to accumulate nitrogen and utilize total nutrients from both preferred and non-preferred leaf tissues. The efficiency and rate of nitrogen accumulation were independently determined for each herbivore on *Prosopis chilensis*. We collected bulk samples of young, intermediate, and mature leaves from three individuals of *P. chilensis* and presented pre-weighed amounts (about 1.0 gm fresh weight) of each tissue type to the various larvae over a 13-hour nighttime period. Several replicates of each experiment were carried out on different nights. In the morning after such an experiment, we collected, dried, and weighed the larval frass and the leftover leaf material. We then analyzed the frass for total nitrogen using the Kjeldahl technique.

As a control, we dried identical amounts of leaf age class tissues from the same sample plants as fed to the larvae and determined water content and total nitrogen. Using the values obtained from the controls, we could calculate the amount of nitrogen present in a consumed portion of leaf tissue. The difference, then, between the amount of nitrogen in the leaf tissue consumed and that present in the frass of an herbivore constituted the amount accumulated by the larvae during the thirteen-hour test period. We called this value (which essentially measures the *net* accumulation rate) the nitrogen accumulation rate (NAR). The values for the various herbivores on different age class tissues of different hosts are given in Table 4-4. Another measure, the mass utilization rate (MUR) also given in Table 4-4, is simply the difference between the dry weight of leaf material eaten and the dry weight of the frass collected in the morning. Both the nitrogen accumulation rates and the mass utilization rates were converted to efficiency measures by dividing the NAR by the total leaf nitrogen content and the MUR by the total dry weight of leaves eaten and expressing each as a percentage value.

The nitrogen accumulation efficiencies (NAE) and the mass utilization efficiencies (MUE) are also listed in Table 4-4. Our mass utilization efficiency measure is calculated in a manner similar to that of the coefficient of digestibility (C.D.) of Waldbauer (1964) and the values we obtained for our three *Prosopis* species and their herbivores are similar to those of Soo Hoo and Fraenkel (1966), who used Waldbauer's coefficient.

After calculating the rates and efficiencies, we could compare the feeding patterns of the different herbivores with respect to the number of host plant species grazed, their tissue preferences, and their abilities to utilize total biomass and accumulate nitrogen from the three leaf age classes.

Tables 4-3 and 4-4 clearly show that Noctuid A, a specialist on *Prosopis chilensis* (see Table 4-1), preferred the young leaf tissue (Table 4-3) and had significantly higher ($p < 0.05$) nitrogen accumulation and mass utilization rates on young tissues than on either intermediate or mature tissues (Table 4-4), although in terms of mass utilization efficiency there is a significant difference ($p < 0.05$) only between the values for young (31.8) and mature tissues

TABLE 4–4 *Nitrogen Assimilation and Mass Utilization for Insect Herbivores Feeding on Three Age Classes of* Prosopis *Leaves*

Insect Taxon	Leaf Age Class		
	Young	Intermediate	Mature
Prosopis chilensis			
Noctuid A			
NAE *	36.5	25.4	−20.1
MUE *	31.8	23.2	14.1
NAR *	6.5	2.8	−0.3
MUR *	86.4	51.6	5.7
Melipotis bisinuata			
NAE *	48.6	36.7	−6.7
MUE *	53.9	43.2	18.9
NAR *	4.6	3.1	−0.1
MUR *	61.0	75.9	18.5
Semiothisa sp.			
NAE *	21.0	30.6	23.7
MUE *	22.6	26.9	21.5
NAR *	1.3	1.5	0.9
MUR *	24.5	28.0	18.9
Prosopis velutina			
Epicauta arizonica			
NAE *	13.8	11.4	−51.0
MUE *	19.0	27.0	1.6
NAR **	0.98	1.0	−2.7
MUR **	23.4	45.3	2.0

* NAE = % nitrogen accumulated of that consumed.
* MUE = % mass utilized of that consumed.
* NAR = mg nitrogen accumulated / 13 hrs / 1.36–1.5 g fresh larval weight.
* MUR = mg mass (total nutrients) utilized / 13 hrs / 1.36–1.5 g fresh wt larvae.
** NAR = mg nitrogen accumulated / 13 hrs / 8 adult individuals (4 males & 4 females).
** MUR = mg mass utilized / 13 hrs / 8 adult individuals (4 males & 4 females).

☐ = preferred tissue

(14.1). The nitrogen measures, however, indicate that there would be a negative nitrogen balance, and consequently reduced fitness, if the noctuid were forced to feed on mature leaves. The oligophagous herbivore, *Melipotis bisinuata*, had a similar pattern in that it preferred the young leaves of all of its three host species. Like the noctuid, it accumulated nitrogen and extracted total biomass significantly better ($p < 0.05$) from young tissues than from the other two leaf classes and showed a negative nitrogen balance when forced to feed on mature leaves. For this herbivore, the only non-significant pairwise comparison ($p < 0.05$) was in the mass utilization rates on young versus intermediate tissue.

In contrast to the herbivores with only one, or a limited range of hosts, *Semiothisa*, the species with the largest number of hosts (six), had a preference for mature tissue on all of its food plants (Table 4-3). Yet, it had a higher nitrogen accumulation rate on both young and intermediate tissues than on mature tissues of *Prosopis chilensis* (Table 4-4) when forced to eat them in the laboratory. Nitrogen accumulation values for young and intermediate tissues did not differ significantly from one another, but nitrogen accumulation efficiencies on both these tissues did differ significantly from that on mature tissues. Comparisons of mass utilization rates between intermediate and mature leaf tissues showed differences, whereas those between young and intermediate, or young and mature tissues did not, but no pairwise comparisons of the mass utilization efficiencies of this herbivore on leaf age classes showed statistically significant differences. These data indicate that, *in a short-term* experiment, this generalist can assimilate nitrogen and extract total biomass better from young and intermediate leaf tissues of *P. chilensis* than from its preferred, mature tissue—an apparent paradox that will be discussed later in the conclusions.

When comparing the feeding patterns and efficiencies of the herbivores we studied near Andalgalá, it is important to note that, when all the insects were feeding on their preferred tissue, Noctuid A (monophagous) had a rate of nitrogen accumulation on *Prosopis chilensis* that is significantly higher than that of *Melipotis bisinuata*, which has a broader diet spectrum (three host plants), and that both of these species had efficiencies four to five times higher than that of *Semiothisa* sp., a polyphagous species feeding on six plant taxa. However, both the specialist and the oligophagous species on *Prosopis chilensis* had negative nitrogen accumulation rates on mature tissues whereas *Semiothisa* sp. was the only herbivore tested with a positive nitrogen rate on mature tissues—the age class on which it is found in the field.

In Arizona, similar experiments were performed with *Epicauta arizonica*, a specialist herbivore on *Prosopis velutina* (Table 4-1). As shown in Tables 4-3 and 4-4, this insect follows the same pattern as that exhibited by the specialists on *P. chilensis* at our Argentine study site. It prefers the young leaf tissues of its host, and, although it did not accumulate nitrogen significantly better in the laboratory from young leaf tissue than from intermediate tissue, it did exhibit a negative nitrogen balance when forced to feed on mature tissue. Table 4-4 shows that *E. arizonica* has its highest (significant in all pairwise combinations) mass utilization rate and utilization efficiency on intermediate tissues.

CHEMICAL AND PHYSICAL PROPERTIES OF PROSOPIS LEAVES

In order to understand the reasons behind various levels of plant species specificity and tissue preferences, we investigated various chemical and physical properties of the different leaf tissues of the host species of our herbivores. Consequently, we determined nitrogen, total alkaloid and water contents as well as toughness of *Prosopis* leaves to determine their effects, if any, on the observed feeding patterns of the herbivores. All determinations (Tables 4-5, 4-6, 4-7, 4-8) for *Prosopis* species were performed on leaves of the same age classes and/or from the same plants as those used in the laboratory preference studies (Table 4-3). For *P. velutina, P. torquata,* and *P. flexuosa,* measurements were made from two, or all three, of the leaf age classes from each of four individuals. In the case of *P. chilensis,* three leaf age classes from each of eight plants were used. We used air dried tissues for the Kjeldahl nitrogen determinations of all species and analyzed other chemical properties and leaf toughness of *P. velutina* and *P. flexuosa* from material that had been fresh frozen over dry ice in the field and thawed just prior to the taking of measurements. Total alkaloid and water content of *P. chilensis* were determined from fresh material. In all cases, we measured the leaf toughness by the penetometer method (Feeny, 1970) and total alkaloids by titration following the procedure outlined by Bull et al. (1968) assuming an equivalent weight of 140. We considered water content to be the difference in weight of the leaves before and after oven drying at 60°C.

TABLE 4-5 *Distribution of Alkaloids in Leaf Tissues of* Prosopis *Species*

Plant Species	Total alkaloids (% dry weight)[1]			
	Y	I	M	N
Prosopis chilensis	1.8 ± 0.0	1.7 ± 0.9	0.9 ± 0.4	(8)
P. flexuosa	1.6 ± 0.3	no data	0.7 ± 0.1	(4)
P. torquata	0.4 ± 0.1	no data	0.01 ± 0.0	(4)
P. velutina	1.5 ± 0.4	1.3 ± 0.3	1.0 ± 0.4	(4)

1. Y = young leaves; I = intermediate leaves; M = mature leaves. See Figure 4-2 for an illustration of tissue classes. The standard deviation follows each value. N = sample sign.

TABLE 4-6   *The Degree of Correlation Between Physical and Chemical Properties of Leaf Tissues of* Prosopis chilensis *and the Feeding Pattern of* Semiothisa *sp.*

Plant Number	Leaf Tissue[1]	RTP	% N (Dry Wt)	% Leaf Water	% Alkaloid (Dry Wt)	Leaf Toughness
1875	M	100	3.28	57	0.4	11.0
1876	M	99.4	2.83	56	0.7	11.0
1878	I	78.1	3.53	63	1.1	14.0
1881	M	59.4	3.55	57	0.5	16.0
1878	Y	43.8	5.83	66	1.3	4.0
1875	I	42.5	3.82	65	0.7	1.0
1874	Y	41.9	5.13	66	1.2	1.5
1880	M	38.8	3.60	60	1.6	36.0
1876	Y	28.1	5.45	65	1.8	0.1
1881	I	23.1	2.84	60	0.9	1.8
1874	M	18.8	3.59	61	1.2	20.0
1878	M	17.5	3.70	52	1.2	30.0
1876	I	13.8	3.73	68	2.0	4.0
1877	M	13.1	2.66	57	1.0	19.0
1879	M	10.0	3.51	56	1.1	11.0
1877	I	8.1	3.49	65	2.3	4.3
1881	Y	6.9	6.92	64	1.4	1.3
1880	I	6.3	3.10	65	1.7	4.0
1879	I	3.1	3.53	66	1.6	5.0
1877	Y	3.1	5.90	66	3.1	0.1
1874	I	1.9	3.51	67	3.6	10.0
1875	Y	0.0	6.56	68	1.2	0.5
1879	Y	0.0	6.66	66	2.1	0.2
1880	Y	0.0	5.64	68	2.1	0.1
Correlation coefficient with leaf palatability			−0.36	−0.45	−0.62	0.25
r^2			0.13	0.20	0.38	0.06
Significance			NS	<0.05	<0.01	NS

1. Y = young; I = intermediate; M = mature (leaf age classes).
NS = not significant at the $p \leqslant 0.05$ level, F test.
RTP = relative tissue preference as defined on page 67.

At the present time, alkaloids are the only compounds present in the leaves of *Prosopis* species definitely known to be toxic to higher animals, although vertebrates are known to graze *Prosopis* (see Chapters 7, 9, and 10). Thus far, the alkaloids of *Prosopis* leaves that have been characterized are related to adrenaline and amphetamine, both of which can display sympathetomimetic and psychotropic activity in vertebrates (Merc Index of Chemicals and Drugs, 1960). We have quantified total alkaloids of *Prosopis velutina, P. chilensis, P. torquata,* and *P. flexuosa* in young, intermediate (in two of the

TABLE 4-7   *The Degree of Correlation Between Physical and Chemical Properties of Leaf Tissues of* Prosopis chilensis *and the Feeding Pattern of* Melipotis bisinuata

Plant Number	Leaf Tissue[1]	RTP	% N (Dry Wt)	% Leaf Water	% Alkaloid (Dry Wt)	Leaf Toughness
1876	Y	100	5.45	65	1.8	0.1
1881	Y	95.4	6.92	64	1.4	1.3
1875	Y	72.2	6.56	68	1.2	0.5
1879	Y	38.0	6.66	66	2.1	0.2
1876	I	30.8	3.73	68	2.0	4.0
1875	I	28.5	3.82	65	0.7	1.0
1876	M	24.3	2.83	56	0.7	11.0
1881	I	21.7	2.84	60	0.9	1.8
1878	I	21.7	3.53	63	1.1	14.0
1878	Y	19.4	5.83	65	1.3	4.0
1877	M	19.0	2.66	57	1.0	19.0
1881	M	16.7	3.55	57	0.5	16.0
1877	I	12.9	3.49	65	2.3	4.3
1879	I	10.7	3.53	66	1.6	5.0
1875	M	9.9	3.28	57	0.4	11.0
1879	M	9.5	3.51	56	1.1	11.0
1877	Y	8.4	5.90	66	3.1	0.1
1880	Y	7.6	5.64	68	2.1	0.1
1880	I	6.5	3.10	65	1.7	4.0
1874	M	3.4	3.59	61	1.2	20.0
1874	I	2.7	3.51	67	3.6	10.0
1878	M	0.4	3.70	52	1.2	30.0
1874	Y	0.0	5.13	66	1.2	1.5
1880	M	0.0	3.60	60	1.6	36.0
Correlation coefficient with leaf palatability			0.57	0.27	−0.06	−0.43
r^2			0.32	0.07	0.004	0.19
Significance			<0.01	NS	NS	<0.05

1. Y = young; I = intermediate; M = mature (leaf age classes).
NS = not significant at $p \leqslant 0.05$, F test.
RTP = relative tissue preference as defined on page 67.

four cases), and mature leaves (Table 4-5). In all cases, there is a decreasing gradation of alkaloid concentration from the young to old leaves. Thus, in all cases, the youngest leaves contained the highest alkaloid concentration on a dry weight basis. In fact, for all of the Argentine species of *Prosopis* examined, the concentration of alkaloid in the young leaves was significantly higher at the 5 percent level or less than that of other tissues. Only the comparison between young and intermediate tissues of *P. chilensis* showed no statistically significant difference. The alkaloid contents of the young leaves of *P. chilensis,*

TABLE 4-8   *The Degree of Correlation Between Physical and Chemical Properties of Leaf Tissues of* Prosopis velutina *and the Feeding Pattern of* Epicauta arizonica

Plant Number	Leaf Tissue[1]	RTP	% N (Dry Wt)	% Leaf Water	% Alkaloid (Dry Wt)	Leaf Toughness
1	Y	100	7.7	68	1.0	1.6
2	Y	97	6.2	71	2.0	0.9
3	Y	92	6.7	70	1.5	0.1
4	Y	83	6.6	71	1.5	0.8
5	I	63	5.8	67	1.6	1.2
6	I	42	5.4	62	1.3	3.9
7	I	18	4.8	64	1.0	1.6
8	I	13	4.9	63	1.5	1.5
9	M	0.0	4.4	55	0.6	5.5
10	M	0.0	4.0	58	1.6	10.0
11	M	0.0	4.8	63	1.0	12.0
12	M	0.0	4.3	54	0.9	6.0
Correlation coefficient with leaf palatability			0.95	0.88	0.51	-0.72
r^2			0.90	0.77	0.26	0.52
Significance			<0.001	<0.001	NS	<0.001

1. Y = young; I = intermediate; M = mature (leaf age classes).
NS = not significant at $p \leqslant 0.05$, F test.
RTP = relative tissue preference as defined on page 67.

P. flexuosa, and *P. velutina* did not differ significantly from one another but all species showed significant differences ($p \leqslant 0.05$) when compared to *P. torquata*. The difference in the alkaloid content of the intermediate leaf tissues of *P. chilensis* and *P. velutina* was not significant. The mature leaf alkaloid content showed the same significant pattern as the young leaves. We are now determining qualitative alkaloid patterns for various species.

An interesting relationship emerged when we examined the distribution of alkaloids in leaf tissues of *Prosopis* (Table 4-5) in the light of the leaf tissue preferences of the various herbivores (Table 4-3). The two herbivores with the broadest diets, *Semiothisa* and *Oiketicus geyeri* (Table 4-1), prefer the mature leaves of all their host plants, the tissue type with the lowest alkaloid content in the species of *Prosopis* tested. In contrast, there appears to be a positive correlation between alkaloid content and tissue preference for the monophagous and oliophagous species, all of which prefer the high alkaloid containing young leaves of their host plants. These results suggest that alkaloid content may have little effect on the feeding patterns of the specialists, but a deterrent effect on more generalized herbivores.

In order to test if alkaloid content was an important factor in determining leaf tissue preference, the relative tissue preferences of the various herbivores for leaf tissues with measured alkaloid content was tested in the laboratory. The *Prosopis chilensis* tissue preferences exhibited by *Semiothisa* and *Melipotis bisinuata* were correlated by means of linear regression with the leaf alkaloid, total nitrogen, and water content, as well as with leaf toughness.

We presented sprays of several individual plants of *P. chilensis* to *Semiothisa* larvae until we found eight plants that were differentially "preferred" or had distinct larval RTP values. From these eight plants, twenty-four leafy sprays were collected in the field by cutting branches under water. The eight sprays, their stems immersed continuously, were placed inside a cage with sixty *Semiothisa* larvae which were arranged at random on the sprays. The sprays were arranged close together so as to form an artificial "bush" and allow movement of the larvae from spray to spray. Another comparable eight sprays were placed in a different cage with forty-five larvae of *Melipotis bisinuata*. All experiments were performed during January and February 1974. Four similarly chosen and prepared sprays of *Prosopis velutina* were placed in a cage with forty adult *Epicauta arizonica* at Silver Bell in August 1973. At both the study areas, three replicates were run per night on each of three different nights.

In the morning following a night of laboratory grazing, we removed the larvae or beetles from the cages and recorded the fractions of young, intermediate, and mature tissue that had been eaten from each of the eight sprays. The RTP values for the different tissue types were calculated as described above (Table 4-8). Samples of each tissue type of each of the eight sprays were subsequently analyzed for alkaloids, nitrogen, and leaf toughness.

Our observations on the relative tissue preferences of the test plants and tissue types of *Prosopis chilensis* to *Semiothisa* and *Melipotis bisinuata* and of *P. velutina* to *Epicauta arizonica* plus the measurements of leaf toughness, chemical data, and regression statistics are given in Tables 4-6, 4-7, and 4-8. The feeding pattern of *Semiothisa* sp. grazing on *P. chilensis* (Table 4-6) shows that relative tissue preference was negatively and significantly ($p < 0.05$) correlated with both the water content and the alkaloid content of the leaves. The higher negative correlation ($r = -0.62, p < 0.01$) of leaf palatability with alkaloid content than with water content ($r = -0.45, p < 0.05$) is consistent with a deterrent effect of alkaloids on grazing by *Semiothisa*. It is therefore possible that the high leaf alkaloid content in the young leaves is at least partially responsible for its preference for mature leaves of *P. chilensis*. On the other hand, leaf preferences of the oligophagous *Melipotis bisinuata* for *P. chilensis* (Table 4-7) and of the monophagous *Epicauta arizonica* for *P. velutina* (Table 4-8) showed no significant correlation with leaf alkaloid content. In both of these cases leaf palatability was most highly and positively correlated with leaf nitrogen content suggesting that this positive factor overrides any negative effect of alkaloids in determining their feeding patterns on the young tissues.

SUMMARY AND DISCUSSION

During several seasons, we studied five insect species associated with leaf herbivores of *Prosopis* species near Andalgalá and one species at Silver Bell in order to determine the relationships between the degree of specialization on one or a few plant species, the kinds of leaf tissue eaten, and the characteristics of preferred leaves and tissue types. At the Argentine site, two species, Noctuid A and *Brachyphatnus,* were specialists, or monophagous, on *Prosopis chilensis.* A third species, *Melipotis bisinuata,* was oligophagous and the two remaining taxa, *Semiothisa* and *Oiketicus geyeri,* had larvae which were polyphagous. At the Silver Bell site, we studied only one herbivore, *Epicauta arizonica,* which appeared to be a specialist on *P. velutina.* Of the six herbivores studied, four (three specialists and the oligophagous *Melipotis bisinuata*) preferred the youngest leaves. The only two herbivores which preferred mature leaves were the generalists.

We also found that the generalists were more abundant on *Prosopis chilensis,* the only plant taxon with specialist and generalist grazers, than the two specialists (175 observations for *Semiothisa, Oiketicus geyeri,* and *Melipotis bisinuata* versus 112 for the specialists *Brachyphatnus* and Noctuid A). When we extended our records of herbivores beyond the 5 km by 75 m study area, the differences in abundance became more pronounced (735 combined feeding observations for the generalists but only 112 for the specialists). In other words, during the period of our study, the South American species of *Prosopis* with which we worked were grazed predominantly by generalist insects that fed on both *Prosopis* and other plant genera. Grazing records of these polyphagous herbivores constituted 61 percent of our observations on *P. chilensis* and 100 percent of our observations on *P. flexuosa* and *P. torquata.* However, since our observations were limited to a few summers, we can not be sure if this phenomenon is a general one. In addition, we should point out that although generalized at the specific level, all the herbivores were apparently restricted to one plant family, the Leguminosae. In Arizona, our observations were much more restricted and data on similarities or differences in patterns were not collected.

Our studies showed that nitrogen accumulation rates (NAR in Table 4-4) were significantly higher for all herbivores when fed on young leaves than when they fed on mature leaves. Mass utilization rates and efficiencies and nitrogen assimilation efficiencies (MUR, MUE, and NAE in Table 4-4) were all significantly higher for young than for mature leaves for both of the South American specialists, the North American specialist, and the Argentine oligophagous herbivore.

For the generalist *Semiothisa,* a more complicated picture emerged. The nitrogen accumulation rates for this herbivore were signficantly higher when the larvae were forced to feed on young leaves of *Prosopis chilensis* than on the mature leaves that are its natural food preference. This species thus differed from all of the others tested, whether generalists or specialists, in that it did not show the highest accumulation rate on its naturally preferred tissue.

In light of these data we propose the following as a partial explanation of the feeding pattern of *Prosopis* herbivores, and furthermore propose that a preference by specialist herbivores for young leaves versus a preference by more generalized herbivores for mature leaves may be common feeding patterns on plants which show a within-plant distribution of toxins similar to that shown here by *Prosopis* species.

From a nutritional point of view, young leaves of *Prosopis* provide higher levels of nitrogen (and hence protein), water, and probably other nutrients than mature leaves (for other trees, see Dixon, 1973). They also appear to be easier to digest than mature leaves. Everything else being equal, then, we would expect young leaf tissues to be preferred over mature leaves by monophagous, oligophagous, *and* polyphagous herbivores.

However, all other factors are not equal in that the young leaves of *Prosopis* species, although the most nutritious, also contain the highest levels of toxic alkaloids (Table 4-5). Since by definition polyphagous herbivores are less highly coevolved with any given host plant than are more specialized herbivores, we can postulate that the former group have evolved less potent detoxification mechanisms against the toxic system of any given host plant than have the latter, assuming that all host plants do not utilize an identical toxin system.

For monophagous and oligophagous herbivores, toxins may confer little protection to the plant. These herbivores, by feeding on one plant species or a few species within the same genus, have evolved systems capable of ameliorating the effects of the toxins in the young leaf tissues which, due to their high nutrient content, become the preferred tissue. In fact, specialized herbivores in many cases utilize toxic substances in their host plants as location and feeding cues and exhibit an attractive response to these substances (Harborne, 1972). Sequestering of plant toxins by specialist insects for use as defensive or pheromone substances is also a well-known phenomenon (Rothschild, 1973; Wood, 1973).

The net influence of high toxicity and a "better" nutrient resource for young leaves versus low toxicity and a "poorer" nutrient resource for mature leaves on grazing patterns by specialists and generalists should be as follows. The high toxicity of young tissues should have little repulsive effect on the tissue preferences of specialists and in some cases will have attractive effects. Such a situation may apply in the cases of *Melipotis bisinuata* (three host plants), the unidentified noctuid and *Brachyphatnus* sp. (one host plant), and for *Epicauta arizonica* (one host plant). Grazing patterns by generalist herbivores, on the other hand, should be under the opposing influences of the repulsive effects of toxins and the suitability of the leaf tissue as a nutrient resource. The stronger effect of the toxins may result in many cases, in a preference by generalist herbivores for leaf tissues other than those which are the

TABLE 4-9 *Occurrence of Alkaloids[1] in the Leaves of Prosopis Species[2]*

Species	Tyramine	N-methyltyr-amine	Tryptamine	5-hydroxytrypt-amine	β-phenylethyl-amine	Prosopine	Prosopinine	Reference
Prosopis alba	**		**		**			Graziano et al., 1971
P. africana			'			**	**	Saxton in Carman, 1973
P. glandulosa	**	**						Camp and Norvell, 1966
P. juliflora			**	**				Saxton in Carman, 1973

1. *Prosopis* alkaloids are sometimes referred to as amines because the nitrogen atom is not cyclized in these compounds.
2. The taxonomic placement and geographic distribution of these species are given in the Appendix.

TABLE 4-10 *Within-Plant Distributions of Toxins (% Dry Weight)*[1]

Toxin	Leaf age class			Plant species	Growth form[2]
	Young	Intermediate	Mature		
Decreasing concentrations with tissue age[3]					
Alkaloids	1.5	1.3	1.0	*Prosopis velutina*	WP
	1.8	1.7	0.9	*P. chilensis*	WP
	1.6	–	0.7	*P. flexuosa*	WP
	0.08	–	0.00	*P. torquata*	WP
	0.39	0.34	0.12	*Atropa belladonna* L.	HP
	H	–	L	*Colchium autumnale* L.	HP
	H	I	L	*Solanum tuberosum* L.	HP
	H	–	L	*Lupinus* sp.	HP
	H	–	L	*Conium maculatum* L.	B
	H	–	L	*Cicuta virosa* L.	HP
	H	–	L	*Delphinium barbeyi* Huth.	HP
	H	–	L	*Sarothamnus scoparius* Benth.	WP
	H	–	L	*Datura stramonium* L.	A
Cyanogenic glycosides	4.8	–	2.5	*Heteromeles arbutifolia* Munz	WP
Cyanogenics	0.03	0	0	*Preridium aquilinum* (L.) Kuhn	HP
	0.45	–	L	*Ximenia americana* L.	WP
	H	–	L	*Acacia chiapiensis* Safford	WP
	H	–	L	*A. farnesiana* (L.) Willd.	WP
	H	–	L	*Prunus* sp.	WP
	H	–	L	*Pangium edule* Reinw.	HP
	H	–	L	*Sorghum halepense* Pers.	HP
	H	–	L	*S. vulgare* Pers.	A
	H	–	L	*Plantanus* sp.	WP
	H	–	L	*Lotus corniculatus* L.	HP

TABLE 4-10 (Continued)

Toxin	Leaf age class			Plant species	Growth form[2]
	Young	Intermediate	Mature		
Mustard oil glycosides	0.12	—	0.07	*Brassica* sp.	A
Total Isothiocyanates	0.009	—	0.005	*Brassica oleracea* L.[4]	A
Non-protein amino acids					
Pipecolic acid[5]	H	—	L	*Prosopis velutina*	WP
N-methyl-L-serine[5]	0.8	0.45	0.25	*Dichapetalum cymosum* Engelm.	WP
Hypericin	0.060	0.035	0.026	*Hypericum hirsutum* L.	HP
Increasing concentrations with tissue age					
Nicotine[6]	1.00	2.92	5.36	*Nicotiana tobacum* L.	A
Saponins	0.47	—	0.70	*Dioscorea tokora* Miyabe	HP
	L	—	H	*Agrostemma githago* L.	A
Protoanemonin	L	—	H	*Ranunculus* spp.	A, HP

1. Data from Rhoades and Cates, 1976.
2. Growth form: A = annual; B = biennial; HP = herbaceous perennial; WP = woody perennial.
3. Concentrations: H = highest; I = intermediate; L = lowest.
4. Average of eleven commercial varieties, % fresh weight.
5. Not known definitely to be toxic.
6. Relative dry weight values.

most nutritious. Examples are *Semiothisa* and *Oiketicus geyeri* (six and five host plants respectively), both of which prefer the mature leaf tissues of their host plants.

As noted above, for the hypothesis of leaf herbivore-host plant interactions discussed here to be tenable, it is important that the toxic systems of the host plants differ, since, if they were identical, the generalist species would be as highly coevolved as the specialists. We have investigated the kinds of alkaloids present in the leaves of *Prosopis chilensis, P. flexuosa, P. torquata,* and several other sympatric woody legume taxa. All have very different alkaloid profiles (Cates and Rhoades, in press). Although the *Prosopis* species listed in Table 4-9 are not sympatric, they also show unique alkaloid patterns for each species. Higher content of toxic substances in young leaves is not restricted to *Prosopis,* and most toxic plant substances for which data are available (alkaloids, mustard oil glycosides, and cyanogenic compounds) show a within-plant distribution similar to that found in *Prosopis* (Table 4-10). Other toxic substances such as nicotine show a reverse within-plant distribution. For plants displaying a distribution of toxins like that of *Prosopis,* we expect to observe grazing patterns similar to those in *Prosopis.* We have found that specialized herbivores prefer the young leaves and the more generalized herbivores prefer the mature leaves. A previous generalization in the literature was that all herbivores should prefer young leaves (e.g., Rockwood, 1974). This situation is now seen as a limiting case particularly applicable to specialist herbivores and less applicable or inapplicable to generalist herbivores.

Chapter 5

PROSOPIS
FLOWERS AS A RESOURCE

B.B. Simpson
J.L. Neff
A.R. Moldenke

Compared to most of the perennial plant species at the two study sites, individuals of *Prosopis* are large and, on a per tree basis, provide one of the richest sources of pollen and nectar (Neff, Simpson and Moldenke, 1977). In addition to these abundant resources. *Prosopis velutina, P. chilensis,* and *P. flexuosa* produce flowers at a predictable time since they can bloom more or less independently of yearly fluctuations in rainfall. Moreover, the morphology of *Prosopis* flowers is such that almost all invertebrate floral feeders can harvest both the pollen and nectar. This combination of quantity, consistency, and accessibility has led to the exploitation of *Prosopis* flowers by a great variety of animals. More animals are, in fact, found associated with *Prosopis* flowers than with those of any other plant species (or group of related species) in either of the desert scrub sites.

Most of the visitors are insects, but other animals also utilize the flowers. Some of the insects are obligately dependent upon *Prosopis* flowers while others merely take advantage of them as one of many sources of food. In our comparison of the flowering processes of *Prosopis* and the utilization of their flowers by animals at the North and South American study sites, we have found both marked similarities and notable differences. Many of the similarities in flowering patterns are due to the common ancestry of our species, particularly those in section *Algarobia*. Many insect utilization syndromes, however, appear to be the result of convergent selection pressures. Differences in the pattern of flowering can be ascribed, in part, to the discrepancies in the rainfall regimes (Chapter 1) and to dissimilar evolutionary histories of the genus in the two areas. In examining the role of *Prosopis* as a floral resource we will compare the flowering patterns of the different species of *Prosopis* at our sites. We will then examine the general array of organisms found on the inflorescences, the rewards offered by the flowers, and the foraging patterns of the most important group of flower visitors. Most of the discussion will deal with *Prosopis chilensis, P. flexuosa,* and *P. velutina,* but we will occasionally also make comparisons with *P. torquata,* the screwbean found in Andalgalá.

FLOWERING PATTERNS

An initial examination of the length of time during which *Prosopis chilensis, P. velutina,* and *P. torquata* flowers are available over the course of a year at the two sites (Figures 5-2 and 5-3) indicates rough similarities, but the species do exhibit differences. In Andalgalá, both *P. chilensis* and *P. flexuosa* behave similarly while the screwbean *P. torquata* has a distinct flowering pattern. The two members of section *Algarobia* bloom for a short intense period

FIGURE 5-1. *A flowering branch of* Prosopis velutina, *the velvet mesquite, at the Silver Bell, Arizona, study site. Despite their pale color, the pendant inflorescences stand out against the sky and green foliage.*

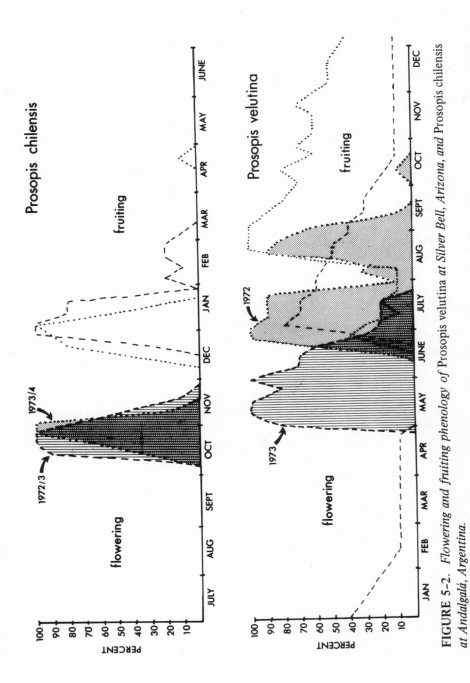

FIGURE 5-2. *Flowering and fruiting phenology of* Prosopis velutina *at Silver Bell, Arizona, and* Prosopis chilensis *at Andalgalá, Argentina.*

early in the spring before the summer rains fall (Figure 1-3) and show no response to the heavy precipitation falling later. Since a phreatophyte is relatively independent of immediate rainfall, it could, presumably, bloom any time during the warm part of the year. However, if a seed has been freed from the endocarp (Figure 6-2) by passing through the digestive tract of an animal (p. 39), successful seed germination and seedling establishment occur only if there is sufficient rainfall to wet the surface layers of soil, cause germination and allow the developing root system to reach a sustaining moisture supply (see Chapter 2). Seeds which remain in the endocarp can remain viable for up to forty years (Martin, 1948), but exposed seeds dry or rot. In the Bolsón, adequate rainfall for germination and establishment occurs only during the summer. Consequently, seeds passed through an animal should be deposited on the ground at the beginning of the rainy season. Flowering therefore occurs one and a half to two months before the rains begin. An accurate timing of flowering is common in many plant species. Such species may use changing day length as the cue for floral initiation. Studies have shown that species of *Prosopis* are sensitive to day length (Peacock and McMillan, 1965) and that the length of day used for a cue depends on the strength of the selective pressure, or the necessary degree of accuracy. In the case of *Prosopis chilensis* and *P. flexuosa,* there appears to be strong selection and flowering occurs only during a very limited period.

In contrast to these two species, the flowering pattern of *Prosopis velutina* is much less restricted (Figure 5-2). As in Argentina, the principal bloom occurs during the spring, but it normally follows a winter rainy season (Figure 5-2). Later in the year, a second bloom, of typically lower intensity than the first, may occur. The duration of the spring flowering of this species is usually greater than that of the two Argentine mesquites. The more flexible flowering pattern is probably due to the differences in the moisture regimes of the two areas. Bailey (1977) has shown that one of the major differences in climate between the Andalgalá and the Silver Bell areas is the period of extreme dryness at Andalgalá before the advent of the summer rains. Although the total amount of precipitation received at the two sites is almost equal, the bimodal rainfall period in Arizona produces an environment that

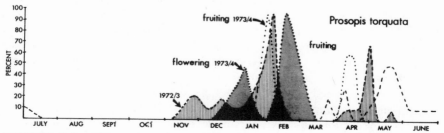

FIGURE 5-3. *Flowering and fruiting phenology of* Prosopis torquata *a non-phreatophytic species of screwbean commonly found near Andalgalá, Argentina.*

never becomes as dry as the early summer in the Bolsón. Thus, for seedlings of plants such as *Prosopis,* there is a comparatively more severe environmental sieve in the Bolsón and seeds that germinate "out of turn" have a lower probability of surviving than those in Arizona. Selection would be expected to be more stringent in the Bolsón for accurately timed seed drop and consequently flowering.

The flowering of the screwbean, *Prosopis torquata,* is unlike that of any of the members of the *Algarobia* group in our areas (Figure 5-3). Structurally quite distinct from true mesquites or algarrobos, this shrubby species occurs on both flats and along washes. Although it was not investigated in detail, the growth pattern of the species suggests that it is much more of a xerophyte than *P. chilensis* or *P. flexuosa.* As is true of most desert xerophytes (Orians and Solbrig, 1977) this species blooms primarily after a significant rainfall although individuals in riparian habitats may flower sporadically throughout the year. As xerophytes the seedlings of *P. torquata* are able to successfully germinate under more severe conditions than those of *P. chilensis* or *P. flexuosa* and the flowering time is not controlled by the timing of seed drop.

During the course of a year, a mature individual of *Prosopis chilensis* or *P. flexuosa* is estimated to produce about 10,000 inflorescences and a weedy, multibranched plant of *P. velutina* (Figure 1-10) about 6,000 (Simpson, unpubl. data). An average plant of *P. torquata* appears to produce about 1500 inflorescences per year (Simpson, unpubl. data). The average length of the inflorescences is also species specific, although within the *Algarobia* group, the number of florets per given length of rachis appears to be quite constant (Solbrig and Cantino, 1975). Inflorescences of *P. torquata* are less compact and have many fewer flowers than the other *Prosopis* species with which it grows (Table 5-1). Despite these morphological differences, the following sequence of development of an inflorescence appears to be similar in all species of *Prosopis* studied.

The inflorescences with small, compact green flower buds appear singly or in clusters of up to eighteen per leaf axil (Figure 5-4), but in members of section *Algarobia,* two or three are most common. The florets along the rachis normally open sequentially from the base to the tip (Figure 5-4). The styles appear first, and seem to be receptive before the flower opens. A day or two later, the petals spread slightly, exposing the anther tips, each surmounted by a gland (Figure 5-5). During the next few days, the petals expand and the anthers uncurl and straighten. Each anther sac splits longitudinally and exposes a mass of pale yellow pollen. The entire process, from the appearance of the first stigmas until the drying of the last flower on an inflorescence, occurs over a period of three days to one week. An individual flower is fully open and presenting pollen and nectar for one day. Consequently, for members of the section *Algarobia,* about one-third of the 150 to 270 flowers, or about 50–90 florets, are fresh on any one day. On an inflorescence of *Prosopis torquata,* about 15 to 20 flowers are available per day (Table 5-1).

TABLE 5-1 *Floral Characters of* Prosopis *Species*

Species	Inflorescence length (mm)	Flowers per inflorescence	Nectar per flower (μl) [1]	Sugar concentration	Sugar per flower (mg)	Total sugar per inflorescence (mg)	Sugar per inflorescence per day (mg)	Pollen per flower (mg) [1]	Total pollen per inflorescence (mg)	Pollen per inflorescence per day (mg)	Mean number of fruits per inflorescence
NORTH AMERICA											
P. velutina	78.1	263	.07 $(20, 10^{-3})$	66%	.046	12.15	4.05	.095 $(13, 10^{-5})$	24.99	8.32	2.148
SOUTH AMERICA											
P. chilensis	82.8	279*				12.89**	4.30**		26.50**	8.83**	1.65
P. flexuosa	62.1	209*				9.66**	3.22**		19.85**	6.618**	1.66
P. torquata	46.6	52.7	.35 $(21, .007)$	24%	.084	4.43	1.48	.178 $(20, 10^{-6})$	9.38	3.13	2.88

1. The first value in parenthesis is sample size, the second is the variance.
*Estimated from the lengths of the inflorescences and the ratios of fls/cm of inflorescence calculated for *P. velutina*.
**Values obtained by utilizing values measured for *P. velutina*.

FIGURE 5-4. *Flowering inflorescences of the velvet mesquite* (Prosopis velutina) *showing the sequential opening of the flowers from the base to the tip of the spike. Flowers at the bottom of the inflorescence are fully open; those nearer the top have just the styles exerted; at the tip, closed buds only.*

In most species of *Prosopis*, the pale greenish-white inflorescences might seem inconspicuous as attraction units to insects and give a superficial impression of an anemophilous catkin. Considerable amounts of pollen may be shaken from the inflorescences, particularly as they dry, and various *Prosopis* species have been listed as major hay fever plants in the American southwest,

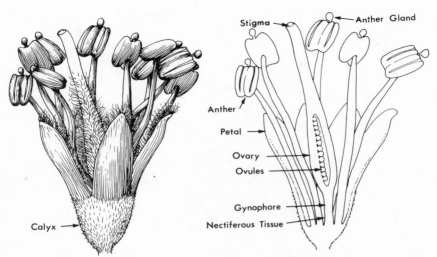

FIGURE 5-5. *A flower of* Prosopis. *On the left, an entire flower showing the reduced calyx, free petals, hairy inner petal surface. The function of the anther gland is unknown. On the right, a schematic cross-section of a flower shows the arrangement of the ovules in the ovary, the gynophore on which the ovary is borne and the zone of tissue that produces nectar. The anther slits and the cup shaped stigma can be seen in both drawings. About 15 times normal size.*

Hawaii, South Africa (in the latter two, mesquite has been introduced) and Argentina (Wodehouse, 1971). Despite the hue of the inflorescences and potentially wind borne pollen, there is no evidence to suggest that wind pollination is anything other than a facultative event. On the contrary, the copious nectar, scented flowers, abundance of insect visitors, and the production of the inflorescences after the leaves have developed (most wind pollinated broad-leaved trees bloom before they leaf since leaves tend to interfere with pollen movement) all suggest that the primary mode of pollination is by invertebrate transport. The sweet odor of the flowers undoubtedly serves as an attractant for many insects, but its precise role in long distance attraction is unknown. Although not colorful, the pendant inflorescences contrast with the green foliage (Figure 5-1) and probably serve as part of the visual cues for insect attraction. In addition, the pollen of *Prosopis* reflects ultraviolet light and anthers with pollen appear as bright dots to insects such as bees that can see short wavelengths of light. We have frequently observed foraging bees to distinguish between pollen rich and pollen poor inflorescences from short distances without landing and physically probing the flowers.

In contrast to the flowers of *Prosopis chilensis*, *P. flexuosa*, and *P. velutina*, the florets of *P. torquata* are bright yellow and during the height of the blooming season, individual shrubs appear as solid masses of gold or yellow.

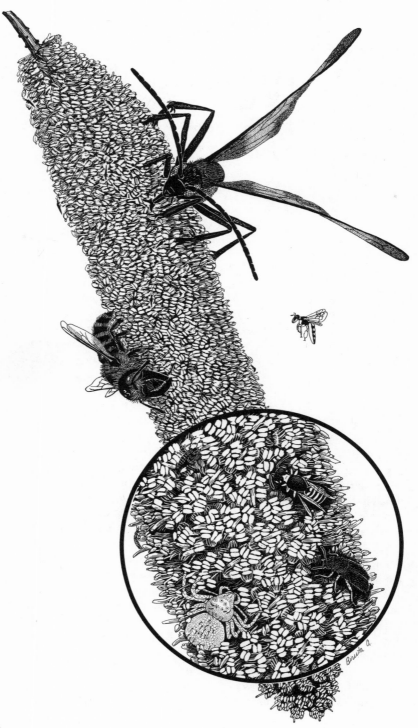

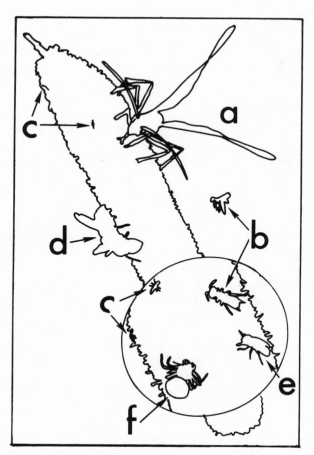

FIGURE 5-6. *Life on a typical* Prosopis velutina *inflorescence near Tucson, Arizona. It is not uncommon to see this many or even more species of Arthropods on a single inflorescence at one time. The inflorescence and insects are about three times normal size and the magnified circle six times natural size.* *a. Tarantula hawk* (Pepsis sp.). *b. Female panurgine bee* (Perdita ashmeadi vierecki). *c. Thirps of various species. Population of these insects sometimes reach into the hundreds on a single inflorescence. d. Female leaf-cutter bee* (Chalicodoma chilopsidis). *e. Melyrid flower beetle* (Trichochrous varius). *f. Crab spider* (Misumenops sp.).

MAJOR GROUPS OF FLOWER VISITORS

During the period they are available, the flowers of a *Prosopis* tree can enter into the lives of literally hundreds of species of desert scrub animals. Figure 5-6 is representative of the fauna on a fresh inflorescence of a mesquite. The inflorescences can serve as hunting sites for carnivores feeding on the many small arthropods that are, in turn, feeding on the flowers. Small ambush predators such as crab spiders (Figure 5-6f) (Thomisidae) are commonly found on inflorescences well camouflaged by their own pale coloration. More infrequently, other vertebrate predators such as mantids (Mantidae), assassin bugs (Reduviidae), or anthocorid bugs (Anthocoridae) patrol the inflorescences. Robber flies (Asilidae) may be found perched on nearby branches searching for invertebrate prey. Insectivorous birds such as hummingbirds or flycatchers (in Argentina) may congregate near mesquite trees in full flower and hawk the flying insects attracted to the inflorescences.

Many insects, usually those species which also feed on *Prosopis,* utilize the inflorescences as mating sites. Adults of many species of seed beetles (Bruchidae), whose larvae will later bore into and destroy the seeds (Chapter 8) in the pods, often feed and mate on the flowers. Other beetles, various flies, moths, butterflies, wasps, and some Hemiptera all occasionally use the flowering spikes as places for mating. The inflorescences serve both as an attractive unit and a perch for those species that mate while resting. The insects most commonly observed using the flowers as mating sites are bees which mate in the air while buzzing around the inflorescences.

Tiny bees of the subfamily Panurginae (Andrenidae), males of which are typically smaller than the females (Figure 5-6b) often fly from flower to flower in copula as the females continue foraging. Male bees of genera such as *Anthidium* or *Protoxaea* (Figure 5-9e) frequently establish territories within a limited portion of the canopy. The males hover within this area, darting out to meet approaching females or any insect or object of approximately the same size. In the majority of bees, however, the males move rapidly around the exterior of the canopy, quickly approaching many inflorescences but usually landing only when females are present. The males may even attempt to grab a female while she is on the inflorescence. Larger bees move in circuits that can encompass several shrubs or trees while the smallest species hover in swarms over clusters of inflorescences, shifting direction in an apparently random manner. Although at first glance such erratic flight behavior might appear to be energetically wasteful, it may be a mechanism for avoiding visual predators such as robber flies. Repeated attempted matings by the males of all species are undoubtedly highly disruptive to efficient pollen collecting by females, since a female usually leaves an inflorescence upon the approach of a male or immediately after attempted (usually unsuccessful) copulation. One side effect of this male-female interaction might be to enhance cross-pollination of *Prosopis* by causing females to visit more distant flowers than they would if left undisturbed.

An analysis of the variety and abundance of the major groups of insects we have collected on the flowers of *Prosopis* in the area near Tucson and around Andalgalá (Neff, unpubl. data) shows that there are no shared species but that the flower feeding faunas of *Prosopis* in the two areas are similar in terms of the major groups that visit the flowers. In both regions, Diptera,

Coleoptera, Lepidoptera, and Hymenoptera are well represented. Likewise, the dominant group of insects in the Monte and the Sonoran Deserts are bees, principally solitary bees. In the southwestern United States alone, over 160 species of solitary bees have been recorded visiting the flowers of *Prosopis* (Moldenke and Neff, 1974; Neff, unpubl. data).

REWARD STRUCTURE

The inflorescences of *Prosopis* are known to contain an unusually high number of secondary compounds (Carman, 1973), many of which have been presumed to play a role in anti-herbivore protection (Chapter 6; Whittaker and Feeny, 1971). Nevertheless floral parts often serve as food. A variety of birds feed on the buds of young inflorescences as a secondary food source. Adult beetles (Figure 5-6e). particulary members of the Melyridae. Tenebrionidae, and Scarabaeidae, feed on floral tissues as well as pollen and nectar. Smith and Ueckart (1974) have recorded mean population densities of up to forty-three individuals of thrips (Figure 5-6c) per inflorescence in Texas and their studies suggest that such infestations may play an important role in reducing fruit production by *Prosopis*. Leaf cutting ants (Attini) occasionally strip entire inflorescences of both buds and flowers and carry them away to serve as a substrate in their underground fungal gardens.

A wide variety of Hemiptera directly insert their stylets and suck phloem and plant juices from young buds, ovaries, rachises, and young fruits. Hemipteran attacks can reach such high levels that they have been implicated as one of the major factors in reducing *Prosopis* seed set (Smith and Ueckart, 1974).

The most abundantly utilized food resources of *Prosopis* flowers are, however, pollen and nectar. Nectar is produced to attract insects which may then serve as inadvertent pollen vectors as they forage from inflorescence to inflorescence and plant to plant. For many insects, it is the primary source of moisture and carbohydrate. In *Prosopis,* nectar is produced by glandular tissue at the base of the style (Figure 5-5). During the day of its life, a floret of *P. velutina* will produce an average of 0.07 μl of nectar containing about 0.46 mg of sugar (Table 5-1; Simpson, 1977, for methods of measuring pollen and nectar). Analyses of the nectar of *P. velutina* at Silver Bell indicates that in this species, a glucose and fructose mixture predominates with lesser amounts of sucrose (Neff, unpubl. data). This finding is in keeping with the deduction of Percival (1961) that glucose and fructose dominated nectars are characteristic of plants with exposed nectar supplies and "unsophisticated" pollination systems. Compared to the 0.6 mg of sugar produced by each flower of *Cercidium microphyllum,* this is a small amount. Yet, since approximately eighty florets are simultaneously producing nectar on a inflorescence of *P. velutina,* the total floral unit produces 2.4 mg per day. On a per visit basis, then, the velvet mesquite potentially produces more sugar than most of the plants in the desert scrub ecosystem. Only the large flowered, generally vertebrate pollinated succulent species such as *Cereus* or *Agave*, produce more total nectar and sugar per floral unit.

The importance of *Prosopis* as a source of nectar has long been recognized by apiarists, because of the quantity of nectar produced and the relative predictability of flowering, even in drought years (Lovell, 1926; Root, 1966). Since the individual flowers are small and without physical barriers barring access to the nectar, the florets are accessible to flower feeders of a wide range of sizes and mouthpart structures. Small bees such as *Perdita* (4-5 mm, Andrenidae, Figure 5-6b) in Arizona or *Pseudiscelis* (3-5 mm, Colletidae) in Argentina and various tiny bombyliid flies may perch on the edge of individual florets while sucking nectar. At the other extreme, large carpenter bees (*Xylocopa* spp., 20-25 mm, Anthophoridae) or tarantula hawks (*Pepsis* spp. 30-40 mm, Pompilidae, Figure 5-6a) climb over the inflorescences rapidly probing many flowers for nectar. In order to amass adequate nectar supplies, these large hymenopterans often walk along the branches within the canopy while searching for untapped inflorescences.

In addition to carbohydrates, adults of various groups of Diptera, Coleoptera, and Hymenoptera need protein and amino acids. For many of these insects, especially bees, pollen provides the only significant source of protein and amino acids. In the case of solitary bees, pollen is the major part of the larval food and must provide enough proper nutrients to ensure development into an adult. As in the case of nectar, with the exception of some Cactaceae, Agavaceae, and a few wind pollinated plants, *Prosopis velutina* produces more pollen per floral unit than most other insect pollinated desert scrub species in North America. While each anther produces but 0.095 mg of pollen (wet weight), each floret has ten such anthers and about eighty flowers on an inflorescence have anthers dehiscing at one time. The total diurnal pollen production from one inflorescence is therefore about 6-9 mg. Unlike the majority of the Mimosoideae whose pollen grains are large sticky tetrads or polyads only infrequently harvested by solitary bees, *Prosopis* pollen is prolate, borne in monads, of medium size (30-34μ) and has a lightly reticulate surface (Figure 5-7). This rather generalized type of pollen is characteristic of insect pollinated plants and requires no specialized morphological or behavioral adaptations for its collection or transport. The pollen is also presented in tiny mounds on each anther face producing a readily gathered supply. The combination of the unspecialized pollen and its accessibility leads to its utilization by a wide variety of bees. When medium or large sized insects land on an inflorescence, the ventral surface is liberally covered by the pollen that protrudes in masses from the flowers. Since all the species of the *Algarobia* group studied appear to be self-incompatible (Simpson, mss.), there is no risk of self-pollination. If the flower visitor subsequently visits flowers of another tree, some of the grains may rub off the undersurface into the stigmatic cup (Figure 5-5). Many of the smaller visitors, however, often move below the level of the anthers and stigmas when foraging for nectar, but females do pick

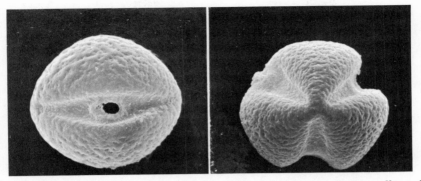

FIGURE 5-7. *Scanning electron microscope photograph of the pollen of* Prosopis velutina. *The grains are very unspecialized, tricolporate, and rugose. All species of the section* Algarobia *have similar grains. The size is about 30 μ.*

up substantial numbers of grains if they are collecting pollen and may thus effect fertilization if they visit another plant.

Since all of the members of the *Algarobia* group that grow sympatrically near Andalgalá bloom at about the same time, have similar flowers, and are visited indiscriminately by bees and wasps, there is ample opportunity for interspecific hybridization. Hybrids between various members of this section have been reported from virtually all areas in which more than one species occurs (Chapter 3).

Unlike the other species of *Prosopis* with which it grows, *Prosopis torquata,* is partially self-compatible, and pollen transfer within an inflorescence could, therefore, result in successful fertilization.

PROSOPIS AND SOLITARY BEES

As we have seen, bees (principally solitary species, e.g., Figure 5-8) are the major group of invertebrates visiting mesquite in both North and South America and are, without doubt, the single most important group of insects effecting pollination. Of the 160 species of solitary bees that have been found associated with *Prosopis* throughout the American southwest, 64 were found visiting the flowers of *P. velutina* during one season at the Silver Bell Bajada in Arizona. At our study site near Andalgalá, 80 species were found visiting the several species of *Prosopis* over the course of two years. Tables 5-2 and 5-3 give the ten most common bees species associated with *Prosopis* at the two sites, along with some of their pertinent attributes. Although many of these species belong to genera which are present at both study sites (e.g., *Colletes, Dialictus, Megachile, Exomalopsis, Melissodes. Centris, Xylocopa*), members of only two genera (*Centris* and *Megachile*) were sufficiently abun-

FIGURE 5-8. *An anthophorid bee,* Svastra bombilans, *collecting pollen on an inflorescence of* Prosopis chilensis *in Andalgalá, Argentina.*

dant on *Prosopis* flowers to be considered major pollinators at both study sites. A very prominent portion of the *Prosopis* bee fauna at both areas is restricted to only one continent. Thus, despite the faunistic similarities of the two regions, similar adaptations in specificity to *Prosopis* or in foraging behavior are in large part the result of convergence and not simply taxonomic heritage. One of the clearest cases of such convergence is the apparent replacement of the numerous tiny, pale species of North American *Perdita* by various small, light colored exomalopsines (*Eremapis, Isomalopsis*), paracolletines (*Bicolletes*), and panurgines (undescribed genus) in Argentina.

Although the majority of the bee species found visiting *Prosopis* at either locality are not restricted to mesquite as either a pollen or nectar host, a significant number in both areas do appear to confine their pollen collecting to *Prosopis* flowers. In North America, the specialists, or oligoleges, are predominantly members of the genera *Perdita* (Andrenidae, Figure 5-9b) and *Colletes* (Colletidae) (see Table 5-4). In North America, *Perdita* is a very large and diverse genus of over 600 species. Its evolutionary history appears to be characterized by oligolecty (pollen host specificity; Linsley, 1958). At least 16 species, primarily in subgenus *Perdita,* can be classified as *Prosopis* oligoleges. Five of these species occur at Silver Bell. These tiny bees (4-5 mm long) are usually the most abundant bees on mesquite throughout the American southwest, and it is common to see thousands of *Perdita* hovering around flowering individuals of *Prosopis* (Cockerell, 1900).

The other genus with a large number of specialists on *Prosopis* is *Colletes.* This is a cosmopolitan genus (absent only from Australia) with mesquite specialists at both of our study areas (Tables 5-4, 5-5). Two species of mesquite-specific *Colletes* occur at Silver Bell. Even when species of *Colletes* not obli-

TABLE 5-2 *Major* Prosopis *Bees of the Silver Bell Bajada*

Taxon	Body length (in mm) ♀	Body length (in mm) ♂	Tongue length (in mm)	Host preferences	Flight Periods III IV V VI VII VIII IX
HALICTIDAE					
Dialictus microlepoides (Ellis)	5–6	4–5	1	Polylectic	
Evylaeus amicus (Cockerell)	7–8	6–8	1	Polylectic	
Nomia tetrazonata Cockerell	9–10	9–10	1.5–2	Polylectic	
ANDRENIDAE					
Perdita ashmeadi vierecki Timberlake	4	3.5	1	Oligolege of Prosopis	
Perdita luciae luciae Cockerell	4–4.5	3.5–4	1	Oligolege of Prosopis	
Perdita obliqua Timberlake	5	4	1	Prosopis is principal pollen host numerous taxa for nectar	
Perdita punctosignata Cockerell	4–5	3.5–4	1	Oligolege of Prosopis	
Perdita stathamae stathamae Timberlake	4–4.5	3.5–4	1	Oligolege of Prosopis	
MEGACHILIDAE					
Megachile (Sayapis) newberryae Cockerell	11–13	11–12	3–3.5	Polylectic	
Megachile (Pseudocentron) sidalceae Cockerell	11–13	10–11	4–5	Polylectic	
Chalicodoma (Chelostomoides) odontostoma Cockerell	9–11	6–9	2	Oligolege of Leguminosae emphasis, Prosopis	
ANTHOPHORIDAE					
Centris rhodopus Cockerell	11–13	11–13	6–7	Polylectic –	
Melissodes paroselae Cockerell	10–12	10–11	3–4	Apparently polylectic predominantly males on Prosopis	

TABLE 5-3 *Major* Prosopis *Bees of the Aldalgalá Area*

Taxon	Body length (in mm) ♀	♂	Tongue length (in mm)	Floral host* preferences	Flight periods X XI XII I II III
COLLETIDAE					
Colletes sp. 1	9–10	8–9	1	Oligolege of *Prosopis*	
Colletes sp. 2	7.5–8.5	7–7.5	1	Oligolege of *Prosopis*	
Bicolletes sp. 1	8–9	6.5–7.5	1–1.5	Oligolege of *Prosopis*	
Oediscelis sp. 1	4–4.5	4	1	Oligolege of *Prosopis*	
ANDRENIDAE					
Liopoeum argentina (Joergenson)	7.5–8.5	7–7.5	2–3	Polylectic	
MEGACHILIDAE					
Megachile (Dasymegachile) sp. 1	13–15	11–13	3.5–4.5	Oligolege of *Prosopis*	
Megachile (Pseudocentron) sp. 2	11–12.5	10–11	4–5	Polylectic	
ANTHOPHORIDAE					
Centris brethesi Schrottky	12–13	12–13	7–8	Polylectic	
Eremapis parvula Oglobin	5	4	1	Oligolege of *Prosopis*	
Exomalopsis sp. 1	6–7	4.5–5.5	2	Polylectic	
Svastrides zebra (Friese)	12–13	12–13	6–7	Polylectic	
Xylocopa splendidula Lepeletier	14–15	14–15	4.5–5.5	Polylectic	

*All considerations of host specificity must be considered tentative in the absence of a data base equivalent to that available for North America. The species assigned numbers are undescribed, not unidentified.

TABLE 5-4 Prosopis *Bee Specialists of Southwestern U.S. and Adjacent Mexico*

Taxon	Body length (mm)		Flight[1] period	Relative abundance
	♂	♀		
COLLETIDAE				
Colletes algarobiae Cockerell	8	9	Apr/Jul–Sep	+
Colletes perileucus Cockerell	10	11.5	Apr–Jun	+
Colletes prosopidis Cockerell	6.5	7.5	Mar–May	++
Hylaeus sejunctus Snelling	3	3.5	Mar	+
ANDRENIDAE**				
Perdita ashmeadi ashmeadi Cockerell	3	4	Mar–May	+
Perdita ashmeadi simulans Timberlake	4	–	Apr–May	+
Perdita ashmeadi vierecki Timberlake	3.5	4	Apr–May	+++
Perdita difficilis Timberlake	3.2	4.5	Apr–May (June)	+++
Perdita duplicata Timberlake	3.5	4	Mar	+
Perdita exclamans Cockerell	2.5	4.5	Apr–May	+++
Perdita genalis genalis Timberlake	4.2	4.7	Mar–May	++
Perdita genalis panamintensis Timberlake	4	4.7	Apr–May	++
Perdita innotata Timberlake	3.5	3.7	Apr	++
Perdita luciae decora Timberlake	3.7	4.5	Apr	++
Perdita luciae luciae Cockerell	3.7	4.5	Apr–May	+++
Perdita macswaini Timberlake	4.2	5	Apr–Jun	++
Perdita mimosae Cockerell	3.5	4.5	Mar–Jul	+
Perdita nigricornis Timberlake	4	–	Apr–Jun	++
Perdita nigronotata Timberlake	4.2	–	Mar	+
Perdita obliqua Timberlake	4	5	May–Aug	+++
Perdita pallidipes Timberlake	4	–	Mar–Apr	+
Perdita prosopidis Timberlake	3	3.8	Apr	+
Perdita punctosignata Cockerell	4	5	Mar–Aug	+++
Perdita sonorensis Cockerell	3.5	4.5	May	+++
Perdita stathamae eluta Timberlake	3.5	4.5	May	+
Perdita stathamae stathamae Timberlake	3.6	4.5	Apr–May	++
Perdita triangulifera Timberlake	3.2	4.5	Apr–Jun	+++
MEGACHILIDAE				
Ashmeadiella prosopidis (Cockerell)	4	5	Mar–Jun	++
Chalicodoma odontostoma Cockerell	7	12	Mar–Jun	++
'*Megachile newberryae* Cockerell	11	12	Mar–Jul	++

1. Flight period refers to the time of the year during which adults of the species are actively flying.
*Probably not obligatedly restricted to *Prosopis* as a pollen host.
**Many of the taxa of *Perdita* are extremely similar morphologically and occasionally their taxonomic status remains uncertain.
+++ Abundant
++ Common
+ Occasional or rare

TABLES 5-5 Prosopis *Bee Specialists of Andalgalá and Adjacent Monte*

Taxon	Body length (mm) ♂	Body length (mm) ♀	Flight[1] period	Relative abundance
COLLETIDAE				
Colletes brethesi Joergensen	8	10	Oct–Nov	+
Colletes sp. 1	8	10	Oct	++
Colletes sp. 2	7	8	Oct	++
Bicolletes sp. 1	7	9	Oct–Nov	+++
Bicolletes wagneri (Vachal)	12	12	Oct–Nov	++
Bicolletes sp. 3	–	9	Nov	+
Bicolletes sp. 4	–	9	Oct	+
*Pseudiscelis sp. 1	3.0	3.2	Oct–Nov	++
*Pseudiscelis sp. 2	–	3.0	Oct–Nov	+
*Pseudiscelis sp. 3	–	3.0	Oct–Nov	+
*Pseudiscelis sp. 4	–	3.0	Nov	+
*Oediscelis sp. 1	4	4.5	Oct–Nov	++
*Oediscelis sp. 2	4.5	5.0	Oct–Nov	++
*Oediscelis sp. 3	4.5	–	Oct	++
ANDRENIDAE				
Genus A sp. 1	4.5	5.0	Oct	+
Genus A sp. 2	4.5	–	Nov	+
Genus B sp. 1	4.5	6.0	Oct	+
MEGACHILIDAE				
Megachile (Dasymegachile) sp. 1	11	13	Oct–Nov	+++
Megachile (Austromegachile) sp. 1	–	17	Oct–Nov	+
Hypanthidoides sp. 1	5.0	5.5	Oct–Nov	+
ANTHOPHORIDAE				
Eremapis parvula Oglobin	4.0	5.0	Oct–Nov (Jan)	+++

1. Flight period refers to the time of the year during which adults of the species are actively flying.
*According to Toro and Michener, 1975, these belong in the genus *Chilicola*.
+++ Abundant
++ Common
+   Occasional or rare

gately associated with *Prosopis* are considered, individuals of this genus tend to be much less abundant than individuals of *Perdita*.

The degree of specificity of *Prosopis* bees in Argentina is hard to determine precisely because of the scarcity of observations over broad geographic areas and by the fact that during the early spring bloom there are few alternate floral hosts available. In the latter case, apparent host specificity at a given site might be an artifact of *Prosopis* being the only available host rather than actual choice. At least fifteen species of bees appear to be restricted to *Prosopis* as a pollen host in Andalgalá (Table 5-5).

Among this group of specialists, *Colletes* is the only genus shared with North America, but the monte species do not appear to be particularly closely related to their North American congeners. With the exception of *Colletes,* all of the South American specialists on mesquite belong to groups of primarily neotropical (i.e., Exomalopsini, Xeromelissinae) or Gondwana (Paracolletini) distribution. Currently, a larger number of species of bees specializing on *Prosopis* are known from North America than from South America. However, nearly twice as many specialists occur at the site near Andalgalá (fifteen) than at the Silver Bell site (seven). When the insect fauna of South America as a whole is known as well as that of North America, it will probably prove to be richer in *Prosopis* specialists. This greater diversity is to be expected from the longer, more complex history of *Prosopis* in South America (see Appendix).

In both our Argentine and Arizonan study areas, the tiny specialist species of bees are often the most abundant visitors of *Prosopis* flowers. However, it is not obvious that they are the most efficacious agents of pollen transfer. As a result of their small size, limited powers of flight, and low individual pollen and nectar requirements, foraging areas of these bees are normally small and unless *Prosopis* densities are high and have virtually contiguous canopies, these bees commonly restrict foraging to a single tree and thus minimize possibilities of cross-pollination. In natural situations along washes in both hemispheres, canopies of *Prosopis* trees are often contiguous, but in areas where *Prosopis* has more recently invaded (e.g., large parts of the Silver Bell area, Figure 1–8), trees are shrubbier, have smaller crowns and are often widely spaced. In many situations, therefore, larger polylectic bees such as *Xylocopa, Megachile,* or *Svastrides* are apparently the dominant pollinators despite their low relative abundance because of their high visitation rates and frequent movement between scattered individuals of *Prosopis.*

Constancy of these effective pollen transport vectors will be maintained while *Prosopis* is in full bloom because it provides greater total rewards than any other species flowering simultaneously in either ecosystem. Our collections show that in Arizona many of the larger generalist bees such as species of *Xylocopa, Megachile, Chalicodoma,* and *Nomia* show considerable constancy on *Prosopis* in the spring. A similar situation is seen in Argentina with the polylectic species of *Svastrides, Centris, Xylocopa,* and *Megachile.* The price a plant must pay for utilizing such a "cornucopia" strategy is that large amounts of energy are invested in nectar and pollen that will be consumed by species that play little or no role as effective pollination agents.

COMPETITION FOR FLORAL REWARDS
AND FORAGING PATTERNS

With such a large number of pollen and nectar feeding bee species attracted to *Prosopis* inflorescences, there is necessarily some competition for the floral resources. We have evidence, although still not complete, that at least some of the bee species minimize competition by spatial and temporal segregation.

ANDALGALA   SILVER BELL

0.5 cm

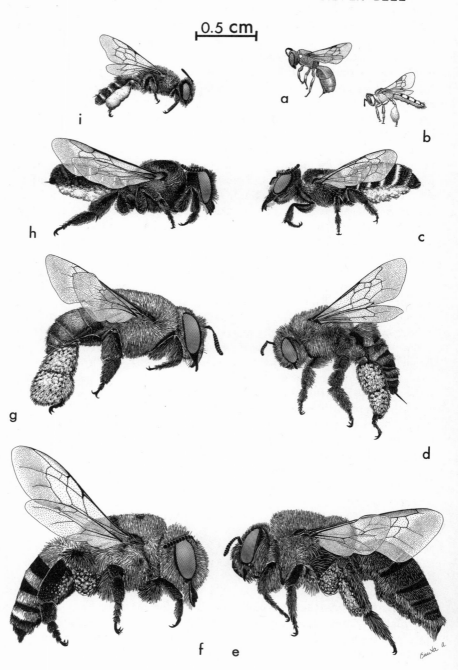

Large melanic bees such as certain species of *Caupolicana, Xylocopa, Megachile,* and *Svastrides* tend to forage primarily on *Prosopis* early in the cool morning and late in the evening when activity of other foragers is at a minimum. Large species in general are usually active earlier in the day than small species of bees in terms of the peak of activity although there may be a total overlap in absolute span of the flight period. A number of the large hymenopterans such as *Pepsis* (Figure 5-6a) or *Xylocopa* also frequently forage in the upper parts of the canopy; but as temperatures rise during the day, the larger species that remain active begin to forage primarily in the shaded interior of the canopy while smaller species remain active around the outer flowers. At night, most bees and other hymenopterans are inactive, but the inflorescences are visited by a variety of moths (Noctuidae and Geometridae) and beetles (Tenebrionidae and Scarabaeidae).

We have also observed a number of seasonal displacements in flight season among bees visiting *Prosopis.* A good example of such a displacement among related species is provided by the different *Perdita* at Silver Bell, all of which are active during the spring bloom. Four *Perdita* taxa. *P. stathamae stathamae. P. ashmeadi vierecki, P. luciae luciae,* and *P. punctosignata flava,* are active only during the early spring. The last two of these species initiate activity earlier in the blooming season, but later the four taxa can be found contemporaneously. A fifth species, *P. obliqua,* is active quite late in the spring and its flight period overlaps only minimally with that of the other four. This species may also have a second emergence following the summer rains when there is a second *Prosopis* bloom.

In Andalgalá, where the members of the *Algarobia* group bloom only once, and for a short period, we would expect that *Prosopis* specialists would have only a matching, single short flight season. One species, visiting *P. chilensis* and *P. flexuosa* in the spring, *Eremapis parvula,* was observed to have a second summer flight period, but during this time it visited *Prosopis torquata.*

Another way in which competition between the many bees utilizing *Prospis* flowers may be mitigated is by the utilization of alternative foraging strategies. For example, members of some species quickly examine many inflorescences and reject those that are not heavily laden with pollen and nectar. When a "suitable" inflorescence is encountered, the bees move over it quickly (Figure 5-8), harvesting only the most accessible pollen and nectar. These

FIGURE 5-9. *Similarities in the collecting apparti of* Prosopis *bees at Silver Bell, Arizona, and Andalgalá, Argentina. The only unique collecting process is found in* Hylaeus episcopalis *(a). This species has no external structures for pollen collecting because it ingests pollen and later regurgitates it. All bees drawn to scale. a.* Hylaeus episcopalis coquilletti. *b.* Perdita punctosignata punctosignata. *c.* Chalicodoma chilopsidis. *d.* Melissodes tristis. *e.* Protoxaea gloriosa. *f.* Caupolicana mendocina. *g.* Centris brethesi. *h.* Megachile cinerea. *i.* Liopoeum argentina.

species appear to be exchanging efficient extraction per foraging bout for a high total yield by rapid but incomplete exploitation of a large number of inflorescences. Examples of this strategy include most of the large, strong flying polylectic (generalist) species such as *Xylocopa, Protoxaea, Caupolicana,* and many Megachilidae. Such a foraging strategy increases the probability of pollen transfer between individuals of *Prosopis* and thus renders these large generalists effective pollinators.

At the opposite extreme from the giant "cream-skimmers" are many of the tiny bees which meticulously gather all available pollen from each anther of each floret. A good example of this strategy is found in the various *Prosopis*-feeding species of *Perdita*. Female bees work each anther with their mouthparts, transferring the pollen to hairs on the ventral surface of the thorax with the forelegs. After working a series of anthers and accumulating a sufficient quantity of pollen, it is transferred to the scopal hairs of the hind leg (Figure 5-9b). This behavior allows tiny bees to utilize inflorescences with flowers which had been previously "skimmed" by another species. This foraging strategy, followed by many generalists as well as specialists, appears to maximize harvesting efficiency but does so by reducing the probability of locating "untapped" flowers on inflorescences.

Behavioral aspects of foraging are often associated with specialized morphological structures for pollen transport. There are numerous examples in the *Prosopis* bees of Silver Bell and the Bolsón de Pipanaco of similarities in these structures. Some examples of bees with similar collecting apparati are shown in Figure 5-9. Members of the Panurginae (e.g., *Perdita,* Figure 5-9b; *Liopoeum,* Figure 5-9i; *Acamptopoeum*) transport pollen by agglutination—a process in which nectar is regurgitated, mixed with pollen and the resultant sticky mass glued to the sparse scopal hairs of the hind legs. Species in the Apidae (*Apis, Bombus, Trigona*) use a similar technique but pack the mass into the corbiculae, specialized structures of the hind tibia. In most other taxa, pollen is usually collected in the dry state and packed into a brush of hairs on the venter of the abdomen (all members of the Megachilidae, Figure 5-9c,h, some Paraolletini, and some Chilicolinae) or among the long, often plumose scopae of the hindlegs (Halictidae, Colletidae, Oxaeidae, and Anthophoridae, Figures 5-9d,g). Species of *Centris* also collect pollen in the dry state, but some of the species that visit *Prosopis* in both areas mix it with floral oils collected from other species of plants. One unique genus, *Hylaeus* (Figure 5-9a), lacks external pollen holding structures and ingests pollen, carrying it to the nest in the honey crop.

Inflorescences of *Prosopis* are normally heavily visited by potential pollen vectors and there is no obvious reason why most flowers should not be pollinated. However, most of the flowers never set fruit (Solbrig and Cantino, 1975). Although as many as thirty fruits have been recorded from a single inflorescence, the mean number per spike is one to two (Table 5-1). Part of this failure of ovarian development may occasionally be due to unsuccessful fertilization or to ovarian damage by floral herbivores, but most of the lack of maturation of the ovaries is probably due to physiological abortion. A successful development of 150–300 large fruits along a 6 cm rachis would be both mechanically and nutritionally impossible. The fact that only a few fruits per inflorescence ever develop and that those which mature are either in a whorl at one place on the rachis (our data) or rather widely spaced suggests that there is suppression of development of unpollinated flowers for a given length of the inflorescence once an ovary has started development. The whorled appearance is presumably due to the fact that a bee could pollinate the receptive flowers in a ring around the inflorescence and these flowers would then begin simultaneous development. None of these flowers could suppress another pollinated at the same time, but all of them would suppress those nearby on the inflorescence. Foraging bees do not limit their movements to a narrow zone on the inflorescences, but of the flowers with which they come into contact, only those in a very restricted area are normally receptive at a given time.

SUMMARY

As would be expected, the abundance of flowers, their copious amounts of nectar and pollen, and the lack of any morphological structures preventing access to these resources has led to the exploitation of *Prosopis* inflorescences by a wide array of animals, particularly bees. Various animals use the flowers as food, consuming only the pollen and/or nectar, or even frequenting the inflorescences only as mating or hunting sites. Most of these animal groups are unrelated in the two study areas.

The predictability with which *Prosopis* blooms in our two desert study areas has been a contributing factor in the specialization of many species of bees for mesquite flowers. Competition among these specialists has led to a variety of temporal, spatial, and mechanical displacements in foraging behavior. Convergent patterns of foraging behavior are found in the bees of the two desert areas.

Despite the high visitation rates, however, most ovaries never begin to mature. It is probable that hormonal inhibition rather than the lack of successful pollination causes this low fruit set relative to the numbers of flowers produced. Still, successful maturation of an ovary does not insure a successful seed crop. Both immature ovaries and ripe pods are subject to predation by a large number of desert scrub vertebrates and insects. While the maturation of fruits constitutes a second step in reproductive cycle of *Prosopis*, it provides at the same time a potential food source for many organisms that has been actively exploited by both animals and man in deserts of North and South America.

Chapter 6

PROSOPIS FRUITS AS A RESOURCE FOR INVERTEBRATES

J.M. Kingsolver

C.D. Johnson

S.R. Swier

A. Teran

Many properties of *Prosopis* fruits and characteristics of their production make them an important resource for numerous animals. As seen in Chapter 1, the timing of the production of the fruit crop of *Prosopis* is predictable compared to that of opportunistic xerophytic species. In addition, the quantity of fruit produced is great. During a four-year study period, samples of fruits from thirty young trees of *Prosopis velutina* with crown diameters of four meters yielded an average of 0.7 kg dry weight of pods per year per tree. These fruits contained an estimated 5,000 seeds (Glendening and Paulsen, 1955). Larger, mature trees with canopies of about six meters were estimated to be capable of producing more than 16 kg of fruits per tree per year, or about 140,000 seeds. Both of these features, a fixed time of fruit maturation and a high yield, are also found in other species of desert trees such as *Acacia* spp. that grow with *Prosopis* in moist microhabitats. However, the fruits of other such species do not have as high a nutritive value as those of *Prosopis*. Chemical analyses of the entire fruit of *P. velutina* (Table 6-1) show that on a moisture free basis, they contain about 16 percent sugar and 12 percent protein. Substantially higher carbohydrate and protein values have been found in other species of mesquite (Table 6-1). In addition, the seeds and fruits of all species of *Prosopis* tested have been found to be non-toxic to either vertebrates or invertebrates. The fruits are readily eaten not only by humans (Chapters 8 and 10) and other vertebrates (Chapter 7), but also in experimental situations by insects such as *Stator pruininus* that have restricted diets (not including *Prosopis*) under natural conditions.

A predictable, abundant, non-toxic, and highly nutritious food source such as *Prosopis* pods naturally attracts numerous herbivores, many of which destroy the seeds. Although it might seem that the fruit syndrome of *Prosopis* only invites predation and would be selectively disadvantageous because of the high seed vulnerability, we shall see that the complex of characters actually represents a compromise that appears to optimize predator escape, seed dispersal and successful seed germination (see Chapters 7 and 9).

FRUIT STRUCTURE

In order to understand how all of the fruit characters act in concert, it is necessary first to understand the structure of the fruit. The fruit of all species of *Prosopis* are in an indehiscent pod (Figure 6-2) or a one-celled "capsule" or "berry." All of the ovules, some of which will eventually mature into seeds, are in a row along one side of the ovary (Figure 5-5). In the section *Algarobia*, the mature fruit has three layers similar to those of the cultivated green bean. The exocarp, or outer layer is thin, relatively soft, and variously colored when

FIGURE 6-1. *Clusters of fruits of* Prosopis velutina *growing at Silver Bell, Arizona. Emergence holes of species of bruchids can be seen as black dots on many of the pods.*

TABLE 6-1   *Nutritional Composition of* Prosopis *Fruits*[1]

| | Total fruit pod | | Seed kernels only | |
Species	% sugar	% protein	% sugar	% protein
P. velutina (Arizona)	16.0	11.9	–	–
P. glandulosa (Texas)	19.1	13.0	3 to 5	61.2
P. glandulosa (Calif.)	31.0	9.5	–	–
P. aff. alba	25 to 28	7 to 11	–	–
P. caldenia	19.8	22.3	–	–

1. North American data from Walton, 1923. Analyses used air-dried fruits. An average dry pod weighed 3.054 gm with a 10% water content. South American data from Burkart, 1943. Air-dried fruits with a 10–12% water content.

ripe (Figure 1–11). In most species of *Prosopis,* the outer surface is smooth (although it is velvety in immature pods of *P. velutina*) and pure or mottled yellow, brown, red, or black. The mesocarp lying just beneath the outer skin is thick and spongy in the species of this section, and contains the major portion of the sugars and starches (Table 6-1). The innermost layer is also relatively thin, but in the mesquites and algarrobos it is hard and stony. Inside this shell is the seed itself (Figure 6-2). Individual oval, brown seeds have circular markings outlining the area from which the embryo usually emerges. In mesquites and algarrobos, as well as the screwbeans, the seeds are oriented end to end, but in some species they are stacked on top of one another (see Appendix).

INVERTEBRATES FEEDING ON *PROSOPIS* FRUITS

The only invertebrates known to use *Prosopis* fruits as a source of food are insects. Most of this consumption results in fruit abortion and/or seed destruction. Thus, these insects can be considered predators or parasites. Most vertebrates which consume *Prosopis* fruits, although destructive in some cases, generally pass the seeds unharmed and thus act more as agents of seed dispersal than as predators. In this chapter, we deal primarily with insect predators and parasites, most of which show an obligate dependency on *Prosopis* fruits, and outline the relationship with vertebrates that serve as dispersal agents. The utilization of *Prosopis* fruits as part of the diet of mammals and some lizards and birds is outlined in the next chapter.

Insects that feed on *Prosopis* can be divided into two groups, those that feed from the outside and those that feed inside the fruits. The former group includes adults and nymphs of some Hemiptera and larvae of Lepidoptera. Internal feeders are generally larvae of Lepidoptera such as *Phalonia* spp. or Coleoptera such as curculionid weevils, long-horned cerambycid beetles, and bruchids. The age of the pod influences, to a degree, the way in which an in-

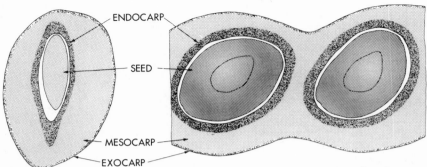

FIGURE 6-2. *Structure of the fruit of* Prosopis *(section* Algarobia*) represented schematically. The exocarp, or "skin," is thin and either solidly colored or mottled. The mesocarp is spongy and usually sweet. The endocarp is hard and stony and similar to a peach pit or an almond shell. At maturity, the seed is free inside a cavity in the endocarp.*

sect can feed. For example, Hemiptera have piercing mouthparts and consequently their feeding is confined to young, tender green fruits. External feeding lepidopteran larvae also usually utilize green and soft pods. On the other hand, the age of the pod is not as crucial for the majority of internal feeders since the seed on which most all of them feed is always soft. For these insects the important step is to get inside the fruit. All internal feeding known is by larval stages. In most cases the adults of these larvae feed on pollen and nectar.

One of the most abundant external *Prosopis* fruit feeders in North America is the hemipteran leaf-footed bug, *Mozena obtusa* (Coreidae; Ueckert, 1973). In Texas, these large insects can be found early in the spring as soon as the buds of *Prosopis glandulosa* begin to break and can remain active until August. The adults pierce the buds and suck them dry, causing them to drop prematurely. Damage to the honey mesquite in Texas was estimated at 50 percent of the potential fruit production (Swenson, 1969). On young fruits of *P. velutina* near Tucson this is also the most frequently encountered hemipteran (Werner and Butler, 1958). Another bug, *Chlorochroa ligata,* similarly pierces the delicate young fruits and causes comparable abortion (Smith and Ueckert, 1974). In a study of *Prosopis flexuosa* near Andalgalá, only 40 of 145 incipient fruits, and in *P. chilensis* only 4 out of an initial 70 fruits, reached maturity (Solbrig and Cantino, 1975). It is possible that insect damage similar to that inflicted by the leaf-footed bug in North America was responsible for much of this fruit abortion in the South American study site although other factors such as water stress, and self-abortion may have also contributed to the mortality. Most lepidopteran larvae that are external feeders simply consume the flowers, immature fruits, and even ripening pods as they crawl along the inflorescence rachises (see Chapter 4).

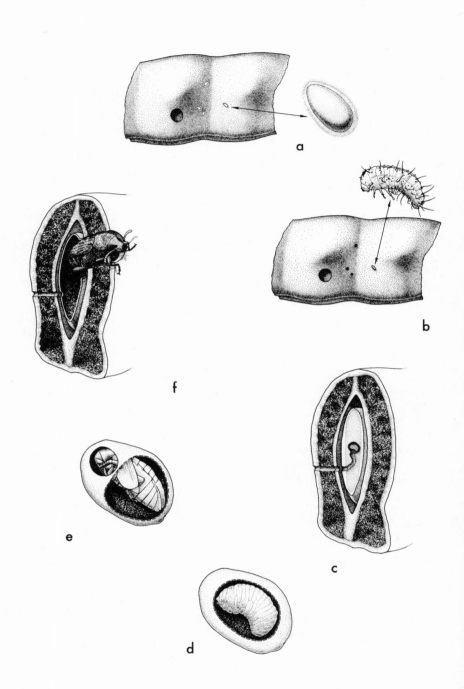

a

b

c

d

e

f

A number of small species of Lepidoptera feed inside the pods of *Prosopis* in the southwestern United States. These taxa include members of the Lycaenidae (*Strymon leda*), Olethreutidae (*Ofatulena* spp.), Pyralididae (*Paramyelois* spp.), Blastodacnidae (*Chaetocampa* sp.), Notodontidae (*Didugua argentilinea*), and Cochylidae (*Phalonia leguminana*) that consume both the seeds and the insides of the maturing pods. Although not reported from the American southwest, long-horn beetles of the genus *Lophopoeum* are known to develop inside the pods of *Prosopis* in Argentina and to consume all of the seeds within a pod while developing. In Arizona, on the other hand, at least three species of weevils (*Apion subornatum, A. ventricosum,* and an unidentified *Microtychius*) use *P. velutina* seeds as food for the larvae developing inside the pods.

BRUCHIDS AND *PROSOPIS*

By far the most numerous (in terms of both species and individuals), most specialized, and best known insects that utilize *Prosopis* fruits as food resources are members of the bruchid family (Kingsolver, 1964, 1967, 1968, 1972). The larvae of most of the estimated 1,000 species of Bruchidae develop in the pods of members of the Leguminosae although some are known to mature in the fruits of other families (Zacher, 1952; Cushman, 1911). A generalized life cycle for a bruchid in a legume pod is shown in Figure 6-3. The female lays her eggs singly or in clusters on the pod surface, in cracks, or even along the edge. In some species of legumes with dehiscent fruits, the female lays eggs directly on the seeds. The larvae hatching from eggs laid on the pod burrow into the fruit and subsequently feed on the tissue of the seed or on both the seed and the fruit pulp (Kannan, 1923). After a succession of molts, the bruchids pupate inside the fruit within the feeding cavity. Species associated with *Prosopis* do not spin cocoons, but often line the nest with frass (Swier, 1974). The adults then chew their way out of the fruit through

FIGURE 6-3. *Generalized life cycle of a bruchid on a fruit of* Prosopis. *a. Eggs glued to the surface of a pod or inserted into old emergence holes of the previous bruchid generation (large circular hole). Fruit with eggs about twice natural size, egg enlarged fifteen times. b. Penetration holes of larvae that have bored into the pod. Pod with holes and egg about 2 x, enlarged first instar larva 15 x. c. Cross-section of a fruit showing the tunnel into the seed bored by the larva, magnified about five times natural size. d. Larva developing inside hollowed seeds (4.5 x). e. Pupa occupying larval feeding chamber. The hole in the seed coat near the head of the pupa is part of the tunnel to the endocarp of the pod cut by the larva before pupation (4.5 x). f. Adult emerging through the precut tunnel (5 x). Figure 6-3a is redrawn from Figure 1 in Pfaffenberger and Johnson (1976).*

the characteristic emergence hole clearly visible on mature fruits (Figure 6-1). Overwintering can occur in the pupal stage in the pod before adult emergence or as an adult, that hides in a sheltered spot. Most adult bruchids eat pollen and nectar and many mate on flowers (Chapter 5).

As shown in Table 6-2, nine genera of Bruchidae are known to use *Prosopis* fruits and seeds. Three of these genera are obligately restricted to *Prosopis* while others use fruits of a variety of legumes often found in association with *Prosopis* such as *Acacia, Cercidium,* etc. In the Western Hemisphere, twenty-nine species of bruchids have been found in *Prosopis* fruits. Of these species, twenty-seven, or 93 percent are obligately restricted to *Prosopis,* but individual species of bruchids may feed on several species of *Prosopis.* Table 6-3 lists the species found in both North America and South America and their *Prosopis* hosts.

TABLE 6-2 *Genera of Bruchidae Found Associated with* Prosopis

Genus	Geographic range	Number of species	Specificity
Algarobius	North America, Venezuela	6	Restricted to *Prosopis.*
Neltumius	SW U.S.A., Northern Mexico	2	Restricted to *Prosopis.*
Mimosestes	North America, West Indies, Venezuela, Hawaii	4	1 species restricted to *Prosopis* other three also attack *Acacia.* Others in genus on *Acacia, Parkinsonia, Cercidium.*
Caryedon	Asia, Africa, 1 tropicopolitan	1	*C. serratus* (Olivier), primarily on *Tamarindus* but attacks *Prosopis* in Hawaii.
Amblycerus	Western Hemisphere	4	Genus has varied food habits. *Prosopis* feeders in West Indies, Venezuela, and Perú
Rhipibruchus	Chile, Argentina	4	Restricted to *Prosopis.*
Scutobruchus	Chile, Argentina, Perú	6	Restricted to *Prosopis.*
Pectinibruchus	Argentina	1	Restricted to *Prosopis.*
Acanthoscelides	Western Hemisphere	1	Genus large with varied food habits.

TABLE 6-3 *Bruchid Species Associated with* Prosopis

Genus and species of Bruchidae	Host species of *Prosopis*
NORTH AMERICA	
ALGAROBIUS	
A. prosopis (Le Conte)	*P. velutina* Wooton, *P. glandulosa* var. *torreyana* (L. Benson) M. C. Johnst., *P. articulata* S. Wats.

TABLE 6-3 (Continued)

Genus and species of Bruchidae	Host species of *Prosopis*
A. bottimeri Kingsolver	*P. glandulosa* var. *glandulosa* Torr., *P. reptans* Benth.
A. sp. A	*Prosopis* sp.
A. sp. B	*P. juliflora* (Swartz) DC
A. sp. C	*Prosopis* sp.
NELTUMIUS	
N. arizonensis (Schaeffer)	*P. velutina* Wooton
N. gibbithorax (Schaeffer)	*P. pubescens* Benth.
MIMOSESTES	
M. protractus (Horn)	*P. velutina* Wooton
M. amicus (Horn)	*P. velutina* Wooton, *Acacia constricta* Benth., *Cercidium floridum* A. Gray, *C. microphyllum* (Torr.) Rose & Johnst.

SOUTH AMERICA

RHIPIBRUCHUS	
R. picturatus (Fahraeus)	*P. alba* Griseb., *P. affinis* Spreng., *P. caldenia* Burk., *P. chilensis* (Mol.) Stuntz, *P. elata* (Burk.) Burk., *P. ferox* Griseb., *P. flexuosa* DC, *P. humilis* Hook. & Arn., *P. nigra* (Griseb.) Hieron., *P. torquata* (Lag.) DC.
R. sp. D.	*P. kuntzei* Harms
R. sp. E.	*P. kuntzei* Harms
R. prosopis Kingsolver	*P. alpataco* Phil., *P. chilensis* (Mol.) Stuntz, *P. juliflora* (Swartz) DC, *P. nigra* (Griseb.) Hieron., *P. sericantha* Hook. & Arn., *P. strombulifera* (Lam.) Benth.
PECTINIBRUCHUS	
P. longiscutus Kingsolver	*P. alba* var. *panta* Burk.
SCUTOBRUCHUS	
S. ceratioborus (Philippi)	*P. alba* Griseb., *P. alpataco* Phil., *P. caldenia* Burk., *P. chilensis* (Mol.) Stuntz, *P. nigra* (Griseb.) Hieron., *P. ruscifolia* Griseb., *P. strombulifera* (Lam.) Benth., *P. torquata* (Lam.) DC
S. sp. F	*P. alba* Griseb., *P. chilensis* (Mol.) Stuntz, *P. hassleri* Harms, *P. nigra* (Griseb.) Hieron, *P. ruscifolia* Griseb., *P. torquata* (Lam.) DC
S. gastoi Kingsolver	*P. tamarugo* Phil.
S. sp. G.	*P. argentina* Burk., *P. alpataco* Phil.
S. sp. H.	*P. nigra* (Griseb.) Hieron.
S. sp. I.	*P. ferox* Griseb.
ACANTHOSCELIDES	
A. longiscutus (Pic)	*P. strombulifera* (Lam.) Benth.
A. sp. J. (Argentina)	*P. caldenia* Burk.
A. sp. K. (Argentina)	*P. nigra* (Griseb.) Hieron.
AMBLYCERUS	
A. piurae (Pierce)	*Prosopis* sp. (Perú)
A. sp. L. (Venezuela)	*P. juliflora* (Swartz) DC

In comparing the bruchid faunas of the more restricted North American Sonoran region and the South American Monte, the most striking feature is the evidence of their independent origins. Not only are the species entirely different, but distinct genera restricted to *Prosopis* have undergone radiations on each of the continents. It therefore appears that although *Prosopis* managed to colonize North America several times, there has been no interchange of the "specialized" bruchid faunas. The only closely related "sister" genera of bruchids in the Americas are *Algarobius* in North America and *Scutobruchus* in South America (Kingsolver, 1968).

MECHANISMS FOR AVOIDING BRUCHID PREDATION

Several recent studies have been concerned with the evolutionary inter-actions of Bruchidae and their hosts (Janzen, 1969, 1971a,b,c, 1972; Forister, 1970; Center and Johnson, 1974). Many legumes seem to follow either of two strategies to avoid seed destruction by bruchids: the formation of toxic seeds or the production of such large quantities of seeds that insect popula-tions cannot consume all of the seeds before at least some of them are re-moved away from the parent plant (Janzen, 1969). Species that have toxic seeds cannot rely on vertebrate dispersal agents or must have seeds that are not inadvertently crushed and digested by vertebrates to any great extent. As seen in Chapter 1, *Prosopis* species have large seeds that provide seedlings with sufficient energy reserves to allow rapid and extended root growth. Such large seeds must be consumed, in part, when the pods are chewed by verte-brates. Consequently, *Prosopis* appears to have adopted the second strategy and produces a very large seed crop that insures that at least some of the seeds will be removed before bruchids can infest them.

In addition to these general strategies of predator avoidance, there are several more specific mechanisms that legumes use to minimize the effects of insect damage. In trees that grow in relatively constant, equable environ-ments, these can include such things as the staggering of the fruiting times of individual plants so that insect populations can not build up, the formation of fleshy seeds that dry out too quickly (if they do not germinate) for a parasite to use them as a larval food, or the maturation of seeds that are so small that they contain too little food to support the complete development of an insect (Janzen, 1969). None of these options are open to *Prosopis* growing in desert environments, as fruiting of all individuals is synchronized to occur at the most advantageous time of year for seed germination, and the seeds must be capable both of withstanding dessication and of providing a good initial energy source for the new seedling (Chapter 1). Several other mechanisms, however, are variously used by species of *Prosopis* (Bridwell, 1918, 1920; Janzen, 1969). These mechanisms include the indehiscence of the pods that excludes all species of bruchids that oviposit on bare seeds, the production of a smooth outer pod surface that precludes species that need cracks or crevices for egg laying or a hairy covering on the pods that hinders the gluing of the eggs to the surface, and the extrusion of a gum that often succeeds in pushing the eggs off the fruit when the pod is pierced by an ovipositing female. The mechanisms for seed protection are not perfect, however, and the seeds of *Prosopis* species are sought, consumed, and destroyed in enormous numbers by bruchid larvae. Some species of *Prosopis* are simultaneously used by up to ten bruchid species; others appear to be used by only one.

BRUCHIDS AND *PROSOPIS* AT SILVER BELL
AND ANDALGALÁ

At Silver Bell, three species of bruchids have been found in the fruits of *Prosopis velutina: Algarobius prosopis, Mimosestes amicus,* and *Neltumius arizonensis.* Fruits of ten trees of *P. velutina* were sampled about a month after the first fruit maturation and were found to have bruchid emergence holes in 11.5 percent of the seeds (Solbrig and Cantino, 1975). Although no cumulative time studies were carried out in the Silver Bell area, observations made on the same species in the Santa Rita Range to the south showed that there was a steady increase in bruchid predation over time (Glendening and Paulsen, 1955). The rise in predation was almost linear, but leveled off after about four months (Figure 6-4), when almost 75 percent of the seeds had been destroyed. These gross figures, however, obscure the differences in feeding behavior of, and amount of this destruction attributable to, the various species of bruchids that simultaneously consume the seeds. A detailed analysis of the interactions between the three species found at Silver Bell and a fourth, *Mimosestes protractus,* in *Prosopis velutina* pods has been made in an area to the north of Tucson in the Verde Valley (Swier, 1974). Among the four species, there are ecological shifts in time of year and fruit stage for oviposition, location of the egg placement, number of generations per year, and condition and site of overwintering individuals. In some cases, it is easy to see how these shifts permit the coexistence of several species of seed predators on the same host, but in others, the persistence of an apparently maladapted species is somewhat perplexing.

Of the four bruchid species studied, the most common was *Algarobius prosopis.* For a given tree, seed mortality caused by the larvae of this species reached 43 percent and constituted 93 percent of all the bruchid predation. This species possesses a number of morphological and behavioral traits which enhances its efficiency in using *Prosopis velutina* fruits. Individuals are capable of successfully ovipositing on pods of differing ages and thus are not restricted as to the time of year for oviposition by age-dependent characteristics of the pod coat. There can be up to three generations per year. Rather than gluing the eggs to the surface of the pod where they may be easily dislodged, females place their eggs in cracks on the surface. The first instar larvae of *A. prosopis* are the only larvae of the four taxa which have

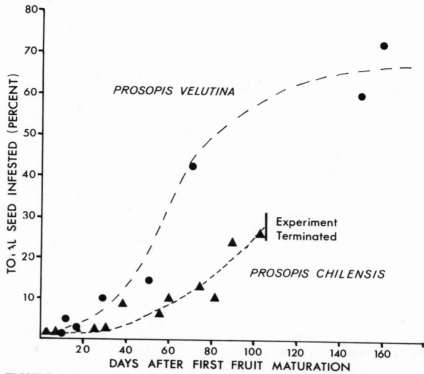

FIGURE 6-4. *Patterns of seed predation of* Prosopis velutina *in Arizona, U.S.A., and of* P. chilensis *in Andalgalá, Argentina, by bruchid (Bruchidae) species.*

well-developed legs that provide mobility and consequently facilitate predator avoidance and also permit an active choice of entry sites into the fruit (Pfaffenberger, 1974). The second most abundant bruchid found in this study, *Mimosestes amicus*, accounted for a lower percentage, 0.7 to 3.38, of the damage to individual trees, or 2 to 3 percent of the total bruchid destruction (Swier, 1974). In contrast to *A. prosopis*, females of this species do not oviposit on very immature pods, but wait until the cotyledons of the enclosed seed are fairly well developed. There are also several generations per year. Eggs are more easily cemented to mature pods than young ones because they are less hairy. Larvae of *M. amicus* have moderately developed legs that allow limited movement on the pod surface. Unlike the other species, *M. amicus* has alternative hosts such as *Acacia* and *Cercidium*. Consequently, although this species is not as abundant on *Prosopis* as *A. prosopis*, it can use different hosts at times of low *Prosopis* fruit production. The third species, *Neltumius arizonensis*, behaves in much the same way as *M. amicus* except that it is restricted to *Prosopis* and is apparently much rarer. Individuals of this species destroyed less than 2.4 percent of the seeds of single trees and accounted for only 0.14 percent of the total bruchid damage. The fact that there is a tendency for this species to use fruits lower in the canopy than those preferred by the other species may explain its low level of occurrence.

The last species, *Mimosestes protractus,* and the one not found in fruits collected at Silver Bell, oviposits on much younger pods than the other three species. By the time *N. arizonensis* and *M. amicus* attack the pod, the larvae of *M. protractus* is in its later instars and is large enough to consume the early instars of the other species boring into the pods. It otherwise appears less well adapted than the other three for using *Prosopis* fruits. Of all the species, it has the most precise oviposition requirements (Figure 6-5) and will lay eggs only along the edge of young pods. This exposed position makes the eggs vulnerable to predation and parasites and allows them to be dislodged easily because of the hairiness and rapid development of the young pods. Larvae are completely legless and must bore into the pods beneath the egg shell. This species also has only one generation per year and, in contrast to the others, overwinters as an adult. It is generally assumed that for bruchids, overwintering in the adult stage carries with it a risk of high mortality.

It has been suggested that most of the specificity and competitive advantages of bruchid beetles on various hosts is due to precise chemical relationships. This does not seem to be the case in *Prosopis*-feeding bruchids. Rather, the kinds of competitive interactions described above seem to limit the numbers of bruchids on a given *Prosopis* species. The lack of chemical specificity is argued for both by the fact that the seeds are not known to be toxic and also because there are numerous examples of shifts to new hosts by bruchids when the *Prosopis* flora is lacking. Within the last hundred years, *Prosopis pallida,* a native of Perú, has been introduced into the Hawaiian Islands as a source of food for livestock (Fosberg, 1966). Inadvertently, several species of bruchids have also been introduced. These include *Mimosestes sallaei* normally found only on *Acacia farnesiana* in North America, *Caryedon serratus,* a species that feeds on numerous legumes including tamarind in its native Asian range, and *Algarobius bottimeri* that feeds only on *P. glandulosa* in Texas (Kingsolver, 1972). In the Hawaiian Islands where these bruchids have been introduced, all use the pods of *P. pallida.* The converse case can also be found. The bruchid *M. amicus,* which has alternate hosts in Arizona, does not attack *P. pallida* in Hawaii. In this new environment, it has been collected only on *Leucaena glauca, Sesbania sesban,* and *Acacia farnesiana* (Hickley, 1960). Finally, in our own collections, *Algarobius prosopis* brought into the laboratory in Washington, D.C., from the Silver Bell Bajada subsequently proceeded to carry out its life cycle in fruits of *P. chilensis* that were stored in the same cabinets as those of *P. velutina.*

All of these data indicate that the fruits of *Prosopis* are readily vulnerable to consumption by bruchid beetles, and that in at least one locality, up to four species of bruchids are known to coexist. This coexistence may be due in part to their attacking the host at different times (Figure 6-5) during the

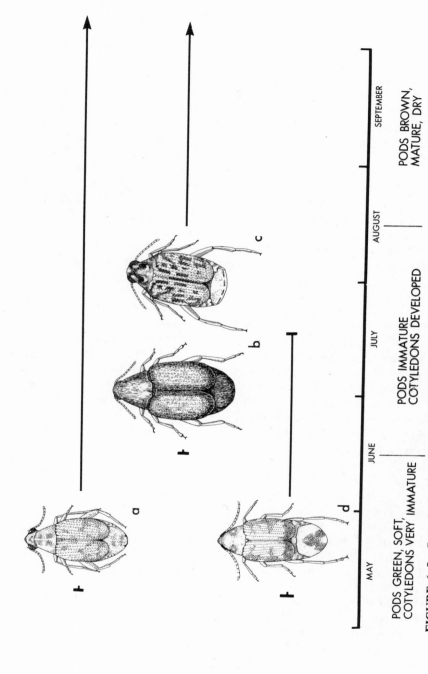

FIGURE 6-5. *Period of oviposition and the stages of* Prosopis *pods on which eggs are laid of four coexisting species of Bruchidae on P. velutina in north-central Arizona. a. Algarobius prosopis. b. Mimosestes amicus. c. Neltumius arizonensis. d. Mimosestes protractus.*

season or by their preferentially ovipositing on pods at different heights in the tree, or by differential ability to use other hosts.

Data from South America, while not strictly comparable, indicate that many of the same types of interactions observed among species on *P. velutina* occur in bruchids infesting algarrobos. Detailed sampling of nine trees of *P. flexuosa* near Andalgalá showed that the total seed destruction (measured by bruchid emergence holes) rose logarithmically over time (Figure 6-4) from the ripening of the first pod to about seventy-five days later. Counts of fruit predation in North America showed a leveling off of parasitism over time. The proportional reduction in predation is due to the fact that unparasitized fruits become harder and harder to find as the season progresses. In South America, we did not witness a similar leveling off of predation because domesticated goats and sheep removed all of the pods which had fallen to the ground. If this had not been the case, we would expect a curve similar to that found in North America.

The final destruction of individual seeds we counted reached 26 percent in *Prosopis flexuosa* and 90 percent in *P. chilensis*. The only species of bruchid found in pods of *P. flexuosa* was *Scutobruchus ceratioborus*. This same species plus two undescribed taxa of *Scutobruchus* and *Rhipibruchus picturatus* were all found in pods of *P. chilensis*. Detailed life history studies of these bruchids similar to those species found infesting *P. velutina* have not been made, but unpublished data (Terán) show that behavioral differences do exist among species of these genera. For example, species of *Scutobruchus* lay eggs individually on green and on mature pods and on pods both in the tree and on the ground. At least one species of *Rhipibruchus* (sp. R.) lays its eggs in partially overlapping "stacks" resembling those of *Mimosestes amicus* in Arizona. Adults of *Rhipibruchus* emerge from the pods of *P. chilensis* before those of *Scutobruchus* with the females emerging before the males in both cases. Species of both genera overwinter as adults that hide in sheltered places such as old pods or fissures in bark. In contrast to the species of bruchids found in North America, the adults of these two genera do not visit flowers for pollen and nectar.

Both at Silver Bell and around Andalgalá, studies show that the amount of damage inflicted by bruchids varies considerably from tree to tree. In the nine trees of *P. flexuosa* monitored (Solbrig and Cantino, 1975) seed damage varied from 0.7 percent in the fallen pods of one tree to 53 percent in those of another. In the study trees of *P. velutina* in Arizona seed destruction varied from 14 percent to 43 percent (Swier, 1974). Studies on *P. velutina* at the Santa Rita Range showed seed destruction as high as 55 percent (Glendening and Paulsen, 1955). The difference in the amount of predation can be ascribed in part to differences between individual trees. The only other significant factors that contributed to variation in the amount of seed destruction were the site of the population, the amount of seed production (fruit production) by a given tree, the direction of the source of bruchids from other trees, and the relative height of the pods in the canopy. However, the significance of the amount of the contribution of each variable differs between bruchids species. As indicated above, for example, there was a significant tendency for most species except *Neltumius arizonensis* to prefer fruits in the upper parts of the canopy.

SUMMARY AND DISCUSSION

Throughout the range of *Prosopis,* there can be no doubt that seed predation, primarily by bruchid beetles, drastically reduces the yield of viable seeds. It is also clear that since successive infestations lead to a logarithmic increase in seed mortality, most seeds that remain near the parent tree will eventually be destroyed (although at a decreasing rate). To "escape" from invertebrate seed predation, *Prosopis* depends on vertebrates that eat the fruits. These animals actively search out the sweet, nutritious pods but do not usually digest the seeds. Undoubtedly, some seeds are crushed when vertebrates chew the pods, but many pass through the digestive tract unharmed and free from bruchids that have been killed by the digestive fluids (Lamprey et al., 1974; Halevy, 1974). In addition, in chewing the pods, these herbivores often remove the impenetrable endocarp and scarify the seeds. The droppings in which the loose seeds are deposited provide a moist and fertile microhabitat for germination. We have found *Prosopis* seeds in the scats of foxes and armadillos, both native animals in the Andalgalá area. Numerous animals in the Sonoran Desert are also known to eat *Prosopis* pods (Chapter 7). Many of these animals consume the pods of other legumes as well and presumably perform the same service as dispersal agents and parasitic exterminators. Animals such as rodents and ants that will eventually eat the seeds often have a tendency to store the fruits (Chapter 7). Frequently, such stored fruits have seeds capable of germinating before they are eaten. The attractiveness of the sweet *Prosopis* pods is indeed evidenced by the avidity with which they are collected by a wide variety of animals. Although the use of mesquite fruits is only part of the food niche of most of these animals, and only one of the many benefits afforded them by trees of *Prosopis,* it is nevertheless a major factor, as we shall see, in maintaining the faunal diversity of the desert regions.

FIGURE 7-1. *A jackrabbit, like many other desert organisms in Arizona, uses the shade of* Prosopis velutina *trees as shelter from the hot summer midday sun.*

Chapter 7

PROSOPIS AS A NICHE COMPONENT

M. A. Mares

F. A. Enders

J. M. Kingsolver

J. L. Neff

B. B. Simpson

Previous chapters have demonstrated the value of many of the characteristics of *Prosopis* such as abundant inflorescences and nutritious fruits for many animals in both desert study areas. In addition, several other features such as the tall, leafy habit of the plants, their complex branching patterns and contiguous distribution along intermittent watercourses or in areas of friable soil and a shallow water table are also important features for many desert scrub organisms. Not surprisingly numerous plants and animals possess particular traits which allow them to exploit the various resources of *Prosopis*. Patterns of exploitation may be of two kinds: those that damage the tree in some way (e.g., leaf-cutting, bark-stripping, etc.) or those that offer some benefit to the trees (dispersal of seeds, aeration of the soil by burrowing, etc.). Even the apparently neutral activity of sitting in the shade of a tree may offer benefits to the plant by concentrating nutrients in the form of excrement under it.

This chapter examines the role of *Prosopis* in relation to plants and animals that utilize trees as a source of shade and/or protection, or food. We will attempt throughout to indicate whether associations existing between mesquite and various organisms are unique to *Prosopis* or whether species use *Prosopis* as merely one of several woody plants in the desert scrub communities. In this way we can assess the relative importance of *Prosopis* in the two ecosystems.

VASCULAR PLANTS*

We can ask several questions when we examine the role of *Prosopis* as a component of the habitat of other plants. For example, does mesquite actually play a particularly important part in the ecologies of other vascular plants? Do the species of *Prosopis* at the two disjunct desert sites fill similar niches? Finally, are any of these associations with other plant taxa unique, or is *Prosopis* merely one of several tree species that can be indiscriminately used by other plants? In order to answer these questions, we will examine numerous ways in which other vascular plants utilize mesquite in the two areas and assess the degree to which these associations are restricted to species of *Prosopis*.

Perhaps the most important feature afforded other plants by individuals of *Prosopis* is that of shelter—shelter from the sun, from soil drought, or, in some cases, from grazing animals (see Chapter 9). Our studies show that species of mesquite which grow along washes and riparian systems provide a distinct microhabitat for annuals and herbaceous perennials under their canopies. In order to illustrate the difference in microhabitat under *Prosopis* canopies and of adjacent exposed sites in both Andalgalá and Silver Bell, we laid out a series of 1 m by 1 m plots under canopies of mesquite trees and on

*B. B. Simpson and J. L. Neff.

exposed areas near the trees. The areas chosen were grazed both beneath the canopies and on exposed areas. We recorded the number of individuals and species of herbaceous plants on each plot. Because there is only one period of rain in Andalgalá, herbaceous plants are present only during the summer. In the Silver Bell Bajada, there are both spring and summer rainfall periods, but for comparative purposes, we emphasized plot samples during the summer season in Arizona.

As shown in Table 7-1, there is a striking difference in both the numbers and kinds of species found under the canopies and in the full sun. Annual grasses, in general, prefer the open, exposed areas. Herbaceous plants with large, mesophytic leaves that would wilt in highly insolated microhabitats grow abundantly under *Prosopis* trees. Some woody perennials such as *Lycium andersoni* at Silver Bell were also observed to grow most frequently along washes under large trees and rarely in the full sun. The preference of large-leaved perennials for sites under trees such as *Prosopis* could be due to their need for the shade provided by the canopy, an increased amount of soil moisture under trees, or both.

In addition to soil moisture, the cooler temperatures under the canopy provide an increased atmospheric moisture relative to surrounding areas. The combination of these factors prevents excessive transpiration and permits growth of the large-leaved herbs. We made no analyses of soil moistures either under trees or at exposed sites, but extensive studies have been made in southeastern Arizona on the effects of *Prosopis velutina* on soil moistures (Glendening and Paulsen, 1955; Parker and Martin, 1952). In these studies, trees of *Prosopis velutina* were killed and the soil profiles around these dead trees used as controls against the soil profiles of live trees (Parker and Martin, 1952). As shown in Figure 7-2, soil moistures were lower under live trees than under dead trees, particularly at lower soil levels. However, it should be noted that these experiments did not test soil moistures under live trees against those in open, exposed areas. Undoubtedly, the branches and twigs of the dead trees provided some shade and the penetration of the soil by the roots probably increased the holding capacity of the soil. Moreover, two other points must be taken into consideration before these results can be extended to our study areas. First, the Santa Rita Experiment Station where these studies were conducted receives more rainfall (34.6 cm versus 29.3 cm) than do our desert sites. As shown in Chapter 2, one consequence of increased soil moisture on the root system of *Prosopis velutina* is an increase in the development of the superficial root system. Our root profile of *Prosopis chilensis* in Andalgalá showed a comparatively poor root mass in the layers of the soil above 46 cm. Second, leaf litter and other material that accumulates under trees of *Prosopis* is probably a major contributor to the retention of soil moisture under the canopy. In the Santa Rita study area, the stands of *Prosopis* investigated are the result of recent invasions and thus trees are relatively young. Consequently, there is a reduced litter layer compared with our sites. It seemed to us, therefore, that in our study areas, both shade and

TABLE 7-1 *Herbaceous Plants Under* Prosopis *versus Open Ground*

Herb taxon	Under canopy number of individuals[a,c]	Exposed number of individuals[b,d]
ARIZONA–Prosopis velutina–SILVER BELL–SUMMER		
Graminae		
Panicum sp. A.	0	1
Chloris virgata Sw.	1	0
Bouteloua aristoides (HBK) Griseb.	397	152
Bouteloua barbata Lag.	10	18
Leptochloa sp. A.	1	0
Eragrostis sp. A.	0	1
TOTAL GRASSES	409	172
Amaranthaceae		
Amaranthus fimbriatus (Torr.) Benth.	10	0
Amaranthus palmeri S. Wats.	23	0
Tidestromia lanuginosa (Nutt.) Standley	37	2
Nyctaginaceae		
Allionia incarnata L.	0	18
Boerhaavia coccinea Mill.	1	0
Boerhaavia erecta L.	1	0
Zygophyllaceae		
Kallstroemia californica (S. Wats.) Vail	31	3
Kallstroemia grandiflora Torr.	16	0
Euphorbiaceae		
Euphorbia sp. A.	90	0
Euphorbia sp. B.	12	0
Euphorbia sp. C.	0	83
Euphorbia sp. D.	0	7
Euphorbia hyssopifolia L.	16	1
Malvaceae		
Sida physocalyx A. Gray	4	0
Sida sp. A.	0	1
Convolvulaceae		
Ipomoea sp. A.	30	0
Solanaceae		
Physalis hederifolia A. Gray	1	0
Acanthaceae		
Ruellia nudiflora (Englm. & Gray) Urban	6	0
Compositae		
Pectis papposa Harv. & Gray	4	49
Ambrosia confertoflora DC	20	0
Acourtia thurberi (Gray) Reveal and King	8	0
TOTAL HERBS	310	164
ANDALGALÁ–Prosopis chilensis–SUMMER		
Graminae		
Setaria viridis (L.) Beauv.	2	0
Aristida adscensionis L.	0	72

TABLE 7-1 **(Continued)**

Herb taxon	Under canopy number of individuals[a,c]	Exposed number of individuals[b,d]
Chloris virgata Sw.	0	16
Bouteloua aristidioides (HBK) Griseb.	0	684
Cottea pappophoroides Knuth	1	1
Eragrostis sp. A.	0	2
TOTAL GRASSES	3	775
Nyctaginaceae		
Allionia incarnata L.	1	6
Boerhaavia coccinea	11	2
Zygophyllaceae		
Tribulus terrestris L.	0	4
Kallstroemia sp. A.	0	2
Euphorbiaceae		
Euphorbia sp. E.	5	0
Convolvulaceae		
Ipomoea sp. B.	1	0
Boraginaceae		
Heliotropium mendocinum Phil.	0	11
Solanaceae		
Solanum sp. A.	59	0
Datura ferox L.	1	0
Compositae		
Verbesina encelioides (Cav.) Benth. & Hook.	104	1
Bidens subalternans DC	17	0
TOTAL HERBS	119	26

a. Average for 3 plots in Arizona.
b. Average for 5 plots in Arizona.
c. Average for 5 plots in Argentina.
d. Average of 10 plots in Argentina.

increased soil moisture relative to fully exposed areas are important factors in providing microhabitats for herbaceous plants. Although *Prosopis* trees are not unique in providing this type of habitat, their dominance along washes in our study areas would make them the most important species serving this function.

Trees of *Prosopis* in both areas also provide a number of structural features important for vascular plants. In addition, several species of parasitic plants in both study areas use either the aerial parts of the plant or the root systems as a source of nutrients. Table 7-2 lists the different ways in which mesquites are used in the two areas and the plant taxa which are associated with them. Mesquite trees can play a passive role as a mechanical shelter or support like almost any other available tree or shrub. Likewise, many of the

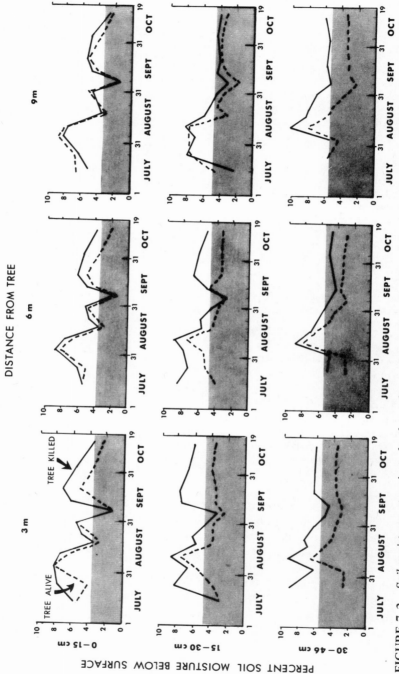

FIGURE 7-2. *Soil moistures at various depths under, and at various distances from, live and killed mesquite trees. The dashed line is the moisture profile during the year for depths and distances under dead trees and the dotted line under live trees. Data from Parker and Martin, 1952.*

parasites and hemiparasites can use tissues of species other than *Prosopis* as food sources.

The only obligate association of any plant taxon with *Prosopis* in the study areas is that between species of algarrobos and *Prosopanche* (Hydnoraceae), a genus of two obligate holoparasites. These bizarre angiosperms lack chlorophyll, and individuals are little more than reproductive organs drawing their complete food source from the roots of their host species. One of the two species of *Prosopanche*, *P. americana* primarily parasitizes *Prosopis* and has been found associated with *P. chilensis*, *P. nigra*, *P. alba*, and *P. flexuosa*, all members of section *Algarobia*, and have only rarely been recorded on other legumes. The second species, *Prospanche bonacinae* occurs on genera as diverse as *Salicornia* (Chenopodiaceae) and *Proustia* (Compositae) but is rarely, if ever, found associated with *Prosopis* or any members of the Leguminosae (Cocucci, 1965).

Flowers of *Prosopanche* are commonly found along washes beneath the canopies of large *Prosopis* trees. The club-shaped protuberances that arise from a complex network of roots are divided into two chambers. The top chamber (Figure 7-3) contains a 3-4 cm tall cylinder composed of fused anthers that produce great quantities of powdery white pollen. Three partially covered openings lead to a lower chamber which houses the broad, flat stigmatic surface (Figure 7-3). Even before the flower opens, weevils of the genus *Oxycorynus*, several species of which appear to be obligately restricted to flowers of *Prosopanche* as their larval host plants, may be found feeding on the exterior of the perianth (Neff, personal observation). As the sepals open, the weevils, and small nitidulid beetles (*Neopadius nitiduloides*), enter the flowers. The weevils continue to feed extensively on the staminal column and tissues lining the inside of the upper chamber of the floral tube. The smaller nitidulids make their way into the lower chamber which lengthens and broadens when the flower opens. Females of *Oxycorynus* oviposit in the holes bored during feeding while the nitidulids lay eggs on the walls of the inner perianth tube. Larvae of both groups feed on floral parenchyma once they hatch. There is a delay of about twenty-four hours after the flowers open before the anthers dehisce. At dehiscence, copious amounts of powdery white pollen are released and often sift into the lower chamber, possibly effecting self-fertilization. Both kinds of beetles are often liberally dusted with pollen, especially as they leave the flowers, and individuals moving to freshly opened *Prosopanche* flowers may thus serve as pollen vectors for cross-pollination. The unusual floral morphology and absence of nectar apparently discourage other potential flower-feeding organisms.

Once fertilized, the ovaries of *Prosopanche* swell and form a large club-shaped fruit (Figure 7-3) containing a white pulp embedded with thousands of tiny, hard seeds. The fermenting fruits have a heavy sweet aroma apparently attractive to mammals such as armadillos and foxes which serve as dispersal agents after they have consumed the fruits and passed the seeds. Small fruit flies breed in the fruits, and ants commonly carry away pieces of pulp and seeds.

TABLE 7-2 Uses of Prosopis by Vascular Plants at Silver Bell and Andalgalá

Use of plant	Part used	Examples of taxa	
		Silver bell	Andalgalá'
Support for climbing	Trunk, branches	Clematis drumondii Torr. & Gray Ipomea spp. Morrenia odorata (H.&A.) Lindl.	Ipomea spp. Curcurbita digitata Gray Sarcostemma crispa Benth.
Holdfast for epiphytes	Branches		Tillandsia xiphioides Ker-Gawl. T. duratii Vis. T. aizoides Mes. T. bryoides Griseb. T. myosura Griseb. T. pedicellata (Mez.) Castell.
Shelter from sun and soil drought	Canopy	See Table 7-1	See Table 7-1
Partial nutrient source (hemiparasites)	Branches	Phoradendron californica Nutt. P. flavescens (Pursh.) Nutt. Strutanthus haenkaena (Presl.) Standley	Phoradendron hieronymi Trel. P. liga (Gill.) Eichl. P. pruinosum Urb. Psittacanthus cuneifolius (R.&P.) Blume
Total nutrient source (parasites)	Roots Roots	Krameria grayi Britt. & Rose None	Ximenia americana L. (?) Prosopanche americana (R. Br.) Baill.

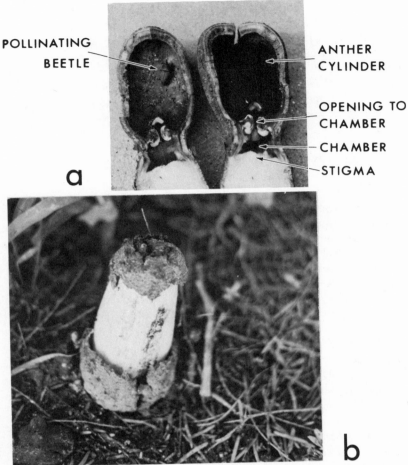

FIGURE 7-3. *Flower and fruit of* Prosopanche americana, *an obligately parasitic angiosperm. Although occasionally found associated with species of other genera,* P. americana *is most commonly found parasitizing the roots of South American* Prosopis *species of the section* Algarobia. *a. The flowers provide no nectar and the pollinating beetles feed on floral parts. Shown is a visiting beetle* Neopadius nitiduloides. *b. Fruit.*

The highly coevolved relationship between *Prosopis. Prosopanche americana,* and the beetles dependent on *Prosopanche* is indicative of a long evolutionary association. The lack of any such intricate utilization of *Prosopis* by a parasitic species in North America suggests a center of origin and longer history of the genus *Prosopis* in South America. In both areas, however, trees of *Prosopis* are generally utilized in much the same way by other higher plants. In most of these cases, *Prosopis* is a representative large desert tree which can provide protection from grazing animals, shade, a moist micro-habitat, a substrate for climbing or perching, and a reliable supply of nutrients for parasitic and hemiparasitic plants. In providing these habitat components desert scrub trees allow an increase in plant density and richness in the community as a whole. These species, which would be rare or absent without the presence of trees and shrubs such as *Prosopis*, in turn contribute to the support of other trophic levels by providing food sources (leaves, flowers, and fruits) for desert scrub animals.

INVERTEBRATES*

Invertebrates form an important, if not always conspicuous, segment of the *Prosopis* community. Despite the number of species of invertebrates known to be associated with mesquite in some way, few studies have been made to determine their dependency on *Prosopis* or their effects on the growth and development of the plants. Arthropods utilize nearly every part of the plant at some stage of its life cycle: some forms feed on living tissue, others feed on dead or dying parts. Still others use *Prosopis* as one of many kinds of wash trees that provide shelter or a hunting ground. Except for those invertebrates we have studied (Chapters 4, 5, and 6), little more can be said about the associations of the majority of invertebrates on *Prosopis* except that they have been collected on, or reared from, plants of mesquite. Other than for our work, relatively few species (generally potential agents for mesquite control) have been studied in detail, and host specificity in most cases can only be surmised from repeated association with *Prosopis* coupled with a lack of association with other plant taxa. *Prosopis* is only one of several tree genera associated with desert washes, and host records of some herbivorous groups indicate that many other trees, principally legumes, can be utilized as effectively. Most invertebrates, particularly wood-boring beetles and seed beetles, belong to groups primarily associated with woody plants. An often-repeated pattern of host genera for such organisms is *Prosopis-Acacia-Parkinsonia-Cercidium* (all members of the Leguminosae).

In addition to the invertebrates that play a role as leaf herbivores (Chapter 4), flower visitors (Chapter 5), or fruit feeders (Chapter 6), we have made comparative field studies only on one other group of arthropods, the spiders.

Quantitative studies of the spider faunas associated with *Prosopis* and other plant genera were made in both study areas by shaking parts of large plants with known volumes or entire small trees and bushes and collecting specimens as they fell. These studies indicated that spiders (as well as coccinellid beetles and lacewings, *Chrysopa* sp., that were inadvertently dislodged) occur in similar densities on both the North and South American species of *Prosopis* when they are in flower. Some preferences for different

*J. M. Kingsolver and F. A. Enders.

plant species by invertebrate predators were found, however. Bugs of the genus *Nabis* (Nabidae) found in both continents and jumping spiders, common at Silver Bell (*Habronattus* and *Phidippus*, Salticidae), prefer species other than *Prosopis*, although they may occur in low abundances on plants of mesquite. In contrast, some web-building and non-web-building spiders are largely restricted to *Prosopis* at both sites.

To confirm their apparent affinity for *Prosopis*, extensive samples were collected from *Prosopis* trees by shaking known volumes over sheets and by net sweeps. Differences in the abundances of non-web-making spiders during the year on plants of *Prosopis* were found in both areas. As shown in Table 7-3, a marked increase in spider abundance at flowering time and a subsequent population decline as flowering ended is evident for *Prosopis*, particularly in South America. In contrast to the pattern of non-web-building spiders, the site selection of web-building spiders was found to be more closely associated with the stiffness of the branches of a plant than with its phenology. For example, in Andalgalá, more web-building spiders such as *Mecynogea* sp., *Metepeira* sp. (Araneidae), *Latrodectus* sp. (Theriidae), *Diquetia catamarquensis* (Diquetidae), and *Dictyna* sp. (Dictynidae) were much more abundant on the stiff-branched *P. torquata*, and *Acacia furcatispina* than on soft-branched species such as *P. chilensis*, *Acacia aroma* or *Larrea* spp. This habitat preference was experimentally verified by the release of spiders on the various plant species.

Overall, the ratio of spiders to insects was lower on *Prosopis* than on *Larrea* or other desert plant species (Chew, 1961; Enders and Bradley, unpubl.) with only *Prosopis torquata* supporting a spider community similar in abundance to that found on plants of mesic areas. At both sites, *Prosopis* had a relatively high abundance of thomisid-like spiders for much of the year and an abundance of clubionid-like spiders only during the period between the initiation of inflorescences and fruit set.

Numerous other small invertebrates use *Prosopis* or other small trees for shelter, nest sites, or as places for nest aggregations, while still different species use *Prosopis* as a primary food source. Various insect taxa have been

TABLE 7-3 *Spiders per Cubic Meter of Foliage of* Prosopis *Species. (Number of Samples in Parenthesis)*

Equivalent month		Arizona	Catamarca	
Arizona	Catamarca	*P. velutina*	*P. chilensis*	*P. torquata*
April	Sept.	3.1 (6)		
May	Oct.		32.6 (1)	
June	Nov.	10.8 (5)		
July	Dec.		8.1 (7)	3.6 (2)
Aug.	Jan.	7.6 (2)		
Sept.	Feb.		2.3 (3)	27.3 (3)

collected in the southwestern United States and northern Argentina that feed on trunks, stems, foliage, roots, or even dead and decaying tissue. Table 7–4 gives an indication of the diversity of insect taxa known to be associated with *Prosopis* in the two areas. Sampling has, of course, been much greater in the American southwest than in Argentina, and the list of taxa in the latter is probably an underestimate.

The types of damage inflicted on trees of *Prosopis* by many of these insects is typical of that incurred by most tree species, whether they are found in mesic or xeric habitats. The number of foliage feeders, however, seems to be considerably lower in desert areas than in more mesic situations perhaps because of the dessication that many such insects would suffer by an exposed feeding behavior. Beetles, in contrast, commonly feed on many parts of *Prosopis* trees in both areas. The adults which are hard shelled and resistant to drying out feed on plant surfaces while immature stages avoid dessication by feeding inside plant parts such as branches, twigs, and seeds.

One of the most spectacular beetles attacking live mesquite branches is *Oncideres rhodosticta*, a gray longhorn beetle obligately restricted to *Prosopis glandulosa* where it occurs in Texas and Arizona. Adults girdle the branches usually smaller than 2.5 cm in diameter, then oviposit terminally to the girdle. The result is a prominant flagging, similar to that produced by oviposition of the seventeen-year cicada. The dying terminal branch segment is then subject to invasion by other beetles such as *Chrysobothris lateralis, C. octocola,* and other buprestids and bostrichids. Ueckart et al. (1971) studied the effects of *Oncideres rhodesticta* on *Prosopis glandulosa* in Texas and found that from 10 to 90 percent of the trees in one area had been attacked. About 40 percent of the branches measuring 0.5 to 2.0 cm in diameter were girdled.

Except for *Onicideres rhodosticta*, some species of Bruchidae (Chapter 6), and a few other taxa (Table 7–4), most invertebrates use *Prosopis* as only one of a number of wash trees. The abundance of mesquite and its phenological pattern, however, automatically make it an integral part of the habitat of many desert invertebrates, particularly those that frequent wash and riparian areas.

MAMMALS*

Is *Prosopis,* per se, important to modern mammals, or do other tall desert trees seem to offer about the same resources? One method of answering this and related questions is to compare the present faunas inhabiting *Prosopis* communities in widely separated areas. If the faunas of these areas are taxonomically unrelated, similar utilization patterns of the animals can be studied, and an assessment made, as to the degree of convergent evolution evident in

*M. A. Mares.

TABLE 7-4 *Insect Associations with* Prosopis *in the Sonoran Desert and the Monte*

Use of plant	Examples of taxa[1]	
	Southwestern United States	Northern Argentina
Shelter		
In leaf debris	Nocturnal spiders	Nocturnal spiders
Loose bark, fissures	Scorpions, mites, spiders, ants, cutworm larvae (*Melipotus* spp.), grasshoppers (*Taeniopoda eques* (Burm.))	Scorpions, mites, spiders, ants
Nest sites	Formicidae (ants)	Formicidae
	Camponotis sayi Emery	*Camponotis mus* Roger
	Megachilidae (leaf cutting bees)	Megachilidae
	Megachile spp., *Chalicodoma* spp.	*Megachile* spp.
Mating aggregations	Chrysomelidae (leaf beetles)	Chrysomelidae
	subfam. Clytrinae	subfam. Clytrinae
Feeding on cell sap	Coccidae	
	Toumeyella mirabilis (Cockerell)	
Rachis of inflorescences and leaves	Miridae	
	Microphylidea prosopidis Knight	
	Phymatopsallus prosopidis Knight	
	Phytocoris lenis Van Duzee	
	Neurocolpus arizonae Knight	
	Orthotylus vigilax Van Duzee	
Flower feeders	Chapter 5	Chapter 5
Fruit feeders	Chapter 6	Chapter 6
Twigs and branches as food	Extensive list including	
	Bostrichidae	
	Buprestidae	Buprestidae
	Cerambycidae	Cerambycidae
	Oncideres rhodosticta Bates	
	Aneflus protensus LeConte	
Dead and dying tissues as food	Buprestidae	Buprestidae
	Chrysobothris lateralis Waterhouse	
	C. octocola LeConte	
	Bostrichidae	
	Xylobiops	
	Xyloblaptus	
	Amphicerus	
	Dendrobiella	
	Isoptera (termites)	
	Kalotermes	
	Procyptotermes	
	Reticulotermes	
Roots	Acrolophidae (Lepidoptera)	Unknown
	Acrolophus spp.	
	Cicadidae (cicadas)	
	Diceroproctus apache Davis	

1. Representative taxa only.

135

the animals that was apparently engendered by the trees. If *Prosopis* is important to the survival of desert vertebrates, we might expect that there exist only a limited number of ways in which the plant resource may be exploited by the animals, and that similar techniques of plant utilization will have evolved in both deserts.

New World desert mammals range in size from the desert shrew (*Notiosorex crawfordi*), which weighs only a few grams, to the jaguar (*Felis onca*), which may exceed 100 kilograms. Within these body size extremes is a rich and diverse fauna of species which, in one way or another, manages to exist in arid and semiarid areas. Many of the larger mammals, particularly carnivores, are wide-ranging species which may be found in a variety of habitat types. Numerous species, especially small mammals (rodents, lagomorphs, and shrews), are much less vagile and may occur in one, or at most a few, habitats. This latter group is particularly interesting from the standpoint of adaptation to particular habitats since many individuals may spend the greater part of their life within one particular habitat type. Mammals utilize *Prosopis* and other desert trees primarily for shade, protection and/or food. Many of the species mentioned in the following discussion are shown in Figures 7-4 and 7-5.

Monte Mammals

In the Monte Desert, the presence of *Prosopis* and associated riparian vegetation influences in a positive manner the number of small mammal species (Table 7-5). Whereas only two or three species of small mammals are found in *Larrea* flats bordering the dry gullies, the forested habitats contain up to six species. Also, small mammals are much more abundant along the gully forests.

The desert cavy, *Microcavia australis*, (Figure 7-4d) and its ecologically similar, less common relative, *Galea musteloides* are typical mammals of *Prosopis* areas in the Monte. In the lowland desert, these medium sized (300 g), diurnal herbivores can be seen scurrying around washes or other forest communities. They forage primarily in the mornings and afternoons, avoiding the desert heat by either sitting in the shade of trees or within their complex burrow systems. Subjectively speaking, cavies seem to prefer *Prosopis* trees (with their globose branching pattern and dense lower branches that reach the ground) for their burrow placement. Also, since they are herbivores, they feed on green vegetation growing within about 100 m of their burrow system. They spend much time foraging in trees, often up to four or more meters off the ground. The tree in which they apparently prefer to climb and forage is *Prosopis chilensis* (in the Andalgalá area). They consume both leaves and seed pods of this tree, and will occasionally strip bark from small, tender branches. I periodically observed (during 1972 and 1973) cavies stripping bark from young *Cercidium* trees, or climbing into *Larrea* bushes at the edge of arroyos, but generally they were found either in, or under, large *Prosopis* trees.

Leaf-eared mice (*Phyllotis griseoflavus*, Figure 7-4e) are also fairly common at times in habitats similar to those frequented by cavies. They are seldom encountered away from riparian areas in the lowland desert, although they are found on the steep, cactus-covered, rocky hillsides which border the northern Monte. These rodents are rather small (80 g), long-tailed, big-eared mice that apparently spend most of their time foraging in trees. I have captured individuals from active cavy systems under *Prosopis* trees. They may also nest in abandoned ovenbird stick-nests, or in nests built in the large cardones, or saguaro-like cacti (*Trichocereus*). (I found only one such nest in a cactus and it was not determined whether or not the nest was built by the mice, or was an abandoned bird nest.) The *rata amarillente maxima* (*Phyllotis griseoflavus*) is nocturnal, and groups of up to six animals may be seen foraging in the same *Prosopis* tree eating leaves (particularly young leaves) and seed pods. They will also eat insects which they encounter on the branches. They climb through an entire *Prosopis* tree using both small and large branches, occasionally squeaking at one another when two meet along the same branch. They are also found in *Acacia* and *Bulnesia* trees, but they appear to be encountered most frequently in mesquite.

The largest non-carnivorous mammal frequenting *Prosopis* areas in the Monte Desert (excluding domestic ungulates) is the Patagonian "hare" (Figure 7-4g) which is not a true hare, but a caviomorph rodent. These animals, which may reach about 15 kilograms, occur throughout much of the lowland Monte in *Larrea* flats or forested areas. They are generally crepuscular (active in early morning and early evening), and appear particularly common in gully forests in the late afternoon, where they may gather in groups of up to six or eight animals. They browse on *Prosopis* leaves and pods from the lower branches.

Tuco-tucos (Ctenomyidae; *Ctenomys fulvus*, Figure 7-4j) are gopher-like rodents that may be found in either creosotebush flats or forested areas. They are herbivorous and commonly feed on *Larrea* bushes that border gullies, but they also occasionally cut small *Cercidium* (palo verde) or *Bulnesia*. Unlike the aforementioned rodents, *Ctenomys* probably have no direct relationship to *Prosopis* for they are essentially completely fossorial and probably do not utilize mesquite for food. They could conceivably feed upon *Prosopis* seedlings, but this has not been observed, and such food habits would only be a small part of a much more extensive plant diet. Other species of mammals frequent *Prosopis* areas in the Monte. Armadillos (*Chaetophractus vellerosus*, Figure 7-4b) and foxes (*Dusicyon griseus*, Figure 7-4h) are both common in washes. The latter has a broad diet which includes *Prosopis* pods (and seeds), as well as insects and vertebrates. The armadillo has a broadly omnivorous diet (Greegor, 1974) and includes *Prosopis* pods as a significant item in its diet. The most common mammal in *Prosopis* areas is

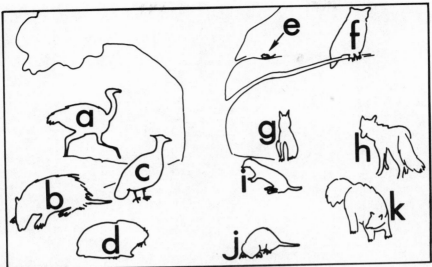

FIGURE 7-4. *Vertebrates commonly associated with* Prosopis *and wash communities near Andalgalá, Argentina. a. Suri* (Rhea americana). *b. Quirquincho chico* (Chaetophractus vellerosus). *c. Perdiz* (Eudromia formosa elegans). *d. Cuis* (Microcavia australis). *e. Rata amarillente maxima* (Phyllotis (Graomys) griseoflavus). *f. Buho* (Bubo virginianus). *g. Mara or Patagonian hare* (Dolichotis patagonum). *h. Zorro gris* (Dusicyon griseus) *i. Huroncito* (Lyncodon patagonicus). *j. Tuco-tuco* (Ctenomys fulvus). *k. Zorrino* (Conepatus chinga).

now the introduced domestic goat (*Ovis capra*). Herds of over a hundred are not uncommon in most gully forests of the northern Monte. They browse extensively on *Prosopis* leaves, twigs, and seed pods. In some localities (see Figure 1-13) a definite browse line is evident where the lower branches of the trees have been eaten to a level of about one and a half meters above the ground. Individual goats will stand on their hind legs and browse as high as they can reach. They may also leap up for vegetation which is just above their reach. One male caught his horns in a fork of a branch after leaping for food and hanged himself. Some goats will even climb onto the lower branches of mesquite to feed. Goat foraging is not limited to *Prosopis* but they certainly spend much time either feeding on the plant, or passing the heat of the day in its shade.

Mammals of North American *Prosopis* Areas

Our studies have shown *Prosopis* communities in North America do not have as marked an effect on the mammal fauna of the area as do those of the Monte (Table 7-6). This statement is not meant to imply that no

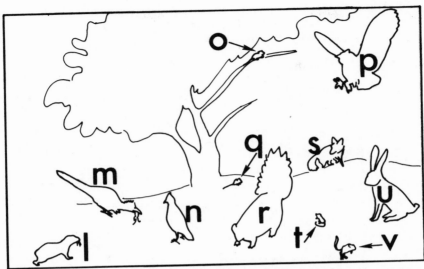

FIGURE 7–5. *Common vertebrates of* Prosopis *communities in North America near Tucson, Arizona. l. Yuma antelope squirrel* (Ammospermophilus harrisi). *m. Roadrunner* (Geococcyx californianus). *n. Gambel's quail* (Lophortyx gambellii). *o. White-throated wood rat or pack rat* (Neotoma albigula). *p. Great horned owl* (Bubo virginianus). *q. Cactus mouse* (Peromyscus eremicus). *r. Hog-nosed skunk* (Conepatus mesoleuceus). *s. Kit fox* (Vulpes microtis). *t. Bailey pocket mouse* (Perognathus baileyi). *u. Black-tailed jackrabbit* (Lepus californicus). *v. Merriam kangaroo rat* (Dipodomys merriami).

species is associated with the riparian community, but rather that the difference in faunas between the tree communities and bordering shrub flats are much smaller than in the Monte. Few Monte mammals are adapted for survival in the desert, whereas in the North American desert systems there are a large number of highly specialized desert forms. Many of these northern taxa may be found in either desert flats or in mesic communities, such as washes, although they are not necessarily found in the same abundance in each habitat. Reynolds and Haskell (1949), for example, found *Perognathus penicillatus* more common in grassy areas where mesquite and cacti were present than in desert grassland with no overstory. Similary, Rosenzweig and Winakur (1969) and Rosenzweig (1973) have shown that *P. penicillatus* prefers areas with taller shrubs. Bateman (1967) also found this species most common in an area of dense mesquite. Nevertheless, I have taken *P. penicillatus* in areas of primarily *Larrea* vegetation, as well as in more forested communities. Monson and Kessler (1940) noted that *Dipodomys spectabilis* frequently placed their burrows under some protective shrub (either *Lycium, Larrea,* or *Prosopis*), although many burrows were constructed in open areas. Schmidt-Nielsen (1964) described how jackrabbits (*Lepus,* Figure 7–5u) uti-

TABLE 7-5 *Monte Small Mammal Faunas: Small Mammal Species of Creosotebush* (Larrea) *Flats and Mesquite* (Prosopis) *Communities in the Northern Monte Desert of Argentina.* [1]

Creosotebush	Mesquite
(I) Andalgalá region (Catamarca, Argentina) Mares, 1973	(M) Andalgalá region (Catamarca, Argentina) Mares, 1973
Eligmodontia typus Meyen (h, 1, 3) *Ctenomys fulvus* Philippi (h, 6, 5) *Dolichotis patagonum* (Zimmermann) (h, 7, 6)	*Phyllotis griseoflavus* (Waterhouse) (o, 4, 2) *Phyllotis* sp. (o, 2, 2) *Galea musteloides* Meyen (h, 6, 2) *Microcavia australis* (Desmarest) (h, 6, 2)
(J) Valle de la Luna (San Juan, Argentina) Mares, 1973	*Dolichotis patagonum* (Zimmermann) *Ctenomys fulvus* Philippi
Eligmodontia typus Meyen *Dolichotus patagonum* (Zimmermann)	
(K) Mendoza city region (Mendoza, Argentina) Mares, 1973	(L) Ñacuñan region (Mendoza, Argentina) Mares, 1973
Eligmodontia typus Meyen *Ctenomys* sp. *Dolichotis patagonum* (Zimmermann) *Marmosa pusilla* (Desmarest) (i, 1, 2)	*Phyllotis griseoflavus* (Waterhouse) *Eligmodontia typus* Meyen *Galea musteloides* Meyen *Microcavia australis* (Desmarest) *Dolichotis patagonum* (Zimmermann) *Lagostomus maximus* (Desmarest) (h, 7, 1) *Ctenomys mendocinus* Philippi (h, 6, 5) *Marmosa pusilla* (Desmarest)

1. Large letters identify the locality of the fauna and are used in Figure 7-6. Small letters refer to food habits and are: granivore (g), herbivore (h), insectivore (i), and omnivore (o). The first number in the parenthesis after food habits refers to body size. Categories of size are: less than 20 g (1), 21–40 g (2), 41–60 g (3), 61–80 g (4), 81–100 g (5) 101–500 g (6), over 500 g (7). The final number in the parenthesis refers to habit. Categories are: ground dwelling (1), scansorial (2), quadrupedal saltation (3), bipedal saltation (4), fossorial (5), cursorial (6).

lize *Prosopis* shade (and other shrubs as well) as an important part of their heat (and water) balance.

Pack rats or wood rats, *Neotoma albigula* (Figure 7-5o), are quite common in gullies where *Prosopis* and other trees are found. Olsen (1973) studied *Neotoma* nesting ecology in southwestern Arizona where *Cercidium* was present. He found that pack rats preferentially built their stick nests under plants which had dense vegetation near ground level. Bateman (1967) also found *Neotoma* common along arroyos with rocky outcrops and dense vegetation. Mares (1973) found white-throated wood rats along washes where

TABLE 7-6 *North American Desert Small Mammal Faunas: Small Mammal Species of Creosotebush (Larrea) Flats and Mesquite (Prosopis) Communities in Desert Regions of North America. Large and Small Letters, Numbers as in Table 7-5.*

Creosotebush	Mesquite
(A) Colorado Desert (California) Ryan, 1968	(B) Colorado Desert (California) Ryan, 1968

(A) Colorado Desert (California) Ryan, 1968

Ammospermophilus leucurus (Merriam)
Spermophilus tereticaudus Baird
Thomomys bottae (Eydoux & Gervais)
Perognathus formosus Merriam
P. fallax Merriam
Dipodomys merriami Mearns
Peromyscus maniculatus (Wagner)
P. eremicus (Baird)
P. crinitus (Merriam)
Neotoma lepida Thomar
Lepus californicus Gray

(C) Yuma area (Arizona) Mares, 1973

Spermophilus tereticaudus Baird
Perognathus amplus Osgood
Dipodomys deserti Stephens (g, 6, 4)
D. merriami Mearns
Lepus californicus Gray
Sylvilagus audubonii (Baird)

(D) Nevada, Bradley and Mauer, 1973

Ammospermophilus leucurus Merriam (o, 6, 2)
Perognathus longimembris Coues (g, 1, 3)
P. formosus Merriam (g, 1, 3)
Dipodomys merriami Mearns
Peromyscus maniculatus (Wagner)
P. eremicus (Baird)
P. crinitus (Merriam) (o, 2, 2)
Onychomys torridus (Coues)
Neotoma lepida Thomar (h, 6, 2)
Lepus californicus Gray
Sylvilagus audubonii (Baird)

(E) Portal area (Arizona) Chew & Chew, 1970

Spermophilus spilosoma (Bennett)
Ammospermophilus harrisi (And. & Bach)
Perognathus baileyi Merriam
Perognathus penicillatus Woodhouse
P. flavus Baird
Dipodomys merriami Mearns
Peromyscus maniculatus (Wagner)
P. eremicus (Baird)
Onychomys torridus (Coues)
Reithrodontomys megalotis (Baird)
Neotoma albigula Hartley
Lepus californicus Gray
Sylvilagus audubonii (Baird)

(F) Mojave Desert (California) Chew and Butterworth, 1964

Ammospermophilus leucurus Merriam
Perognathus longimembris Coues
P. fallax Merriam (g, 1, 3)
Dipodomys merriami Mearns
Peromuscus eremicus (Baird)
Onychomys torridus (Coues)

(B) Colorado Desert (California) Ryan, 1968

Spermophilus tereticaudus Baird
Perognathus pencillatus Woodhouse
Dipodomys merriami Mearns
D. deserti Stephens
Peromyscus maniculatus (Wagner)

(N) Tucson area (Arizona) Mares, 1973

Spermophilus tereticaudus Baird
Perognathus baileyi Merriam
P. penicillatus Woodhouse
Dipodomys merriami Mearns
Reithrodontomys megalotis (Baird) (o, 1, 2)
Peromyscus eremicus (Baird) (o, 2, 2)
Sigmodon hispidus Say & Ord (h, 6, 1)
Neotoma albigula Hartley
Lepus californicus Gray
Sylvilagus audubonii (Baird)

(O) Superior area (Arizona) Bateman, 1967

Pergonathus penicillatus Woodhouse
Dipodomys merriami Mearns
Onychomys torridus (Coues) (i, 2, 2)
Peromyscus eremicus (Baird)
Neotoma albigula Hartley
Lepus californicus Gray
Sylvilagus audubonii (Baird)

(P) Tularosa Basin (New Mexico) Blair, 1943b

Spermophilus spilosoma Bennett
Perognathus flavus Baird
P. penicillatus
Dipodomys ordii Woodhouse
D. merriami Mearns
Peromyscus maniculatus (Wagner)
Onychomys leucogaster (Wied-Neuwied)
Neotoma micropus Baird
Lepus californicus Gray
Sylvilagus audubonii (Baird)

(Q) Colorado Desert (California) Ryan, 1968

Spermophilus tereticadus Baird
Perognathus penicillatus Woodhouse
Lepus californicus Gray
Sylvilagus audubonii (Baird)

(R) Tularosa Basin (New Mexico) Blair, 1941, 1943b

Notiosorex crawfordi (Coues) (i, 1, 1)
Spermophilus variegatus (Erxleben)
S. spilosoma (Bennett)
Cynomys ludovicianus (Ord) (h, 2, 1)
Thomomys baileyi Merriam
Perognathus baileyi Merriam
P. flavus Baird (g, 1, 3)
Dipodomys ordii Woodhouse
D. merriami Mearns

143

TABLE 7–6   (Continued)

Creosotebush	Mesquite
Reithrodontomys megalotis (Baird)	
Neotoma lepida Thomar	*Peromyscus maniculatus* (Wagner)
Lepus californicus Gray	*P. leucopus* (Rafinesque) (o, 2, 2)
	P. eremicus (Baird)
(G) Tularosa Basin (New Mexico) Blair, 1943a	*Onychomys torridus* (Coues)
	O. leucogaster (Wied-Neuwied) (i, 2, 2)
Spermophilus variegatus (Erxleben) (o, 6, 1)	*Reithrodontomys megalotis* (Baird)
Thomomys baileyi Merriam	*Neotoma micropus* Baird
Perognathus penicillatus Woodhouse	*N. albigula* Hartley
Dipodomys merriami Mearns	*Sigmodon hispidus* Say & Ord (h, 6, 1)
D. ordii Woodhouse	*Lepus californicus* Gray
Peromyscus maniculatus (Wagner) (o, 2, 2)	*Sylvilagus audubonii* (Baird)
P. truei (Shufeldt)	
Onychomys torridus Coues	
Neotoma micropus Baird (h, 6, 2)	
Lepus californicus Gray	
Sylvilagus audubonii (Baird)	
(H) Tucson area (Arizona) Mares, 1973	
Spermophilus tereticaudus Baird (o, 6, 2)	
Ammospermophilus harrisi (And. & Bach) (o, 6, 2)	
Perognathus baileyi Merriam (g, 2, 3)	
P. penicillatus	
P. amplus Osgood (g, 1, 3)	
Dipodomys merriami Mearns (g, 3, 4)	
Neotoma albigula Say & Ord (h, 6, 2)	
Lepus californicus Gray (h, 6, 2)	
Sylvilagus audubonii (Baird) (h, 7, 6)	

Prosopis, Acacia, and *Cercidium* were common, as well as in *Larrea* flats (with *Opuntia* present), and on rocky hillsides with cactus.

Occasionally, when trapping in the desert, one encounters a species that is not a usual element of a xeric-adapted fauna. For example, in 1973, I encountered *Reithrodontomys megalotis* and *Sigmodon hispidus* in a mesquite community along a gully near Tucson. Where these species enter the desert, they are most common in more mesic situations.

A number of North American mammals are known to utilize mesquite as food. Reynolds and Glendening (1949, 1950) found that the kangaroo rat (*Dipodomys merriami,* Figure 7–5v) consumed *Prosopis* seeds. On the average, 6.3 percent of the pouches of *D. merriami* contained mesquite beans, with 27.4 percent of the pouches containing beans in July. Mesquite seeds were a preferred food item. The kangaroo rats do not eat all of the seeds but bury many of them in surface caches, where they may later germinate (Reynolds and Glendening, 1950). Alcoze and Zimmerman (1973) showed that, in the mesquite plains of Texas, *Prosopis glandulosa* seeds were among the most important dietary items of *Perognathus hispidus* and *Dipodomys ordii.* Arnold (1942) found that 25.3 percent of the specimens of *Perognathus* (Figure 7–5q,t) which contained food items in their pouches had velvet mesquite seeds, while 11.3 percent of the sample had mesquite flowers, 8.8 percent had leaves, and 7.6 percent contained seed pod sections. Bannertail kangaroo rat burrows (*Dipodomys spectabilis*) contained 2.1 percent by volume of mesquite seeds, which ranked mesquite in the top ten food items encountered (Monson, 1943). Vorhies and Taylor (1922) found that bannertail kangaroo rats stored and ate mesquite beans more than any other non-grass food item. Probably among the most important mammals utilizing *Prosopis,* from the standpoint of quantity of vegetation consumed, are hares (*Lepus*); *Lepus californicus* (Figure 7-5u), the black-tailed jackrabbit, had a diet consisting of 56 percent mesquite (presumably leaves and bark), while for *L. alleni,* the antelope jackrabbit, mesquite formed 36 percent of the diet (Vorhies and Taylor, 1933). As Vorhies and Taylor noted, mesquite was the alternative to green grass. When grass was present, consumption of mesquite was reduced, and the converse was also true. They also described the "browse line" of jackrabbits on mesquite, similar (but lower) to that mentioned above for Monte goats. Packrats are also important *Prosopis* consumers (30.2 percent of their annual diet) and climb throughout the tree while foraging (Vorhies and Taylor, 1940; Bateman, 1967). Goats are not common domesticated animals in the North American study area and domesticated cattle, while avid consumers of *Prosopis* pods, do not browse mesquite branches. The lack of a browse line in Figure 1-9 contrasts with that shown in Figure 1-13.

Obviously *Prosopis* is an important habitat component and is used as a food or place resource by a large number of mammals in North America. Without *Prosopis* we might logically expect a diminution of individuals of a number of species, but probably the overall effect on the mammal fauna would be small. In the Monte however, removing *Prosopis* from the northern lowland deserts would probably greatly reduce the populations of most small mammals. Mesquite, it seems, is a more important plant element near Andalgalá than it is in the North American deserts.

Patterns of Mammalian Diversity

If the small mammal faunas of various *Prosopis* areas are compared (Tables 7-4 and 7-5, it is apparent that the number of total species in each area and the relative species diversity between *Larrea* and *Prosopis* habitats varies greatly. Generally, sites with more rainfall support more species. Lowest species richness (expressed as the number of species per site) is found near Deep Canyon in the Mojave Desert (four species). In Argentina, the xeric Andalgalá site also supports few mammals (six species). Greatest richness in both North and South American areas occurs in semidesert sites. The New Mexico area studied by Blair (1941, 1943a) supports twenty species of small mammals, while the Ñacuñan locality in Mendoza, Argentina (Mares, 1973), has eight species, the highest richness found in any Northern Monte area. The Ñacuñan site is an ecotonal area between arid desert scrub (Monte) and more mesic thorn scrub (Chaco). In physiognomy, it is not unlike Blair's New Mexico area. In general, desert mammal species richness is lower in Argentina than in North America (Mares, 1973).

Certainly the number of species inhabiting an area is important, but it is a simplified measure of diversity. Diversity also includes the relative abundance of species in a habitat (MacArthur and MacArthur, 1961). Other diversity indices may reflect behavioral and/or morphological aspects of a community of mammals, and thus be useful in analyzing a fauna's response to particular environmental characteristics (Fleming, 1973). In the following analyses, data on small mammals from each site (taken from Tables 7-4 and 7-5) were categorized along three major niche parameters: Food (granivore, leaf-eating herbivore, insectivore, omnivore); Body Size (<20g, 21-40g, 41-60g, 61-80g, 81-100g, 101-500g, >500g); and Adaptation, which described where or how animals foraged (ground dwelling, scansorial, quadrupedal saltation, bipedal saltation, fossorial, cursorial). Natural history accounts and other data used in the analyses are from Cabrera and Yepes (1940), Eisenberg (1963), Burt and Grossenheider (1964), and Mares (1973). By using these three niche parameters, one can calculate an index of ecological diversity (Fleming, 1973), which describes the manner in which species in each mammal community are distributed among the various descriptive categories. Since the Shannon-Wiener information theory function is used,

$$H' = - \Sigma p_i \log p_i$$

(H' is the calculated diversity where p_i is the proportion of species found in the ith class) the distribution of species among the various categories is important in determining the maximum diversity value for that particular niche parameter. Thus, six species distributed evenly among six categories would yield a higher H' value than would six species occurring in only one of six categories. The function is thus a probability determination; the more improbable (i.e., unpredictable) a particular occurrence is, the more diversity there exists in that particular system, and vice versa.

Although overall species richness (number per area) influences the calculation of ecological diversity, the effect of the former is moderated somewhat. If a broader range of categories is represented in a numerically inferior fauna, it will be more diverse, ecologically, than a fauna with more species, but with most of these species being in very few categories. As Fleming (1973) noted, the "evenness" component of diversity is important, and is calculated by E= H'/H'$_{max}$. Here E varies from 0 (where all species are in one category) to 1 (where each category has an equal representation of species). H'$_{max}$ is equivalent to the natural logarithm of the number of classes.

A three-dimensional representation of ecological diversity in various desert communities containing *Prosopis* is shown in Figure 7-6, based on information given in Tables 7-4, 7-5. Since the maximum diversity occurs at the upper right-hand corner of the "box" (indicated by an asterisk), obviously neither *Larrea* nor the *Prosopis* wash community in Argentina possesses the ecological diversity evidenced by the majority of North American *Prosopis*

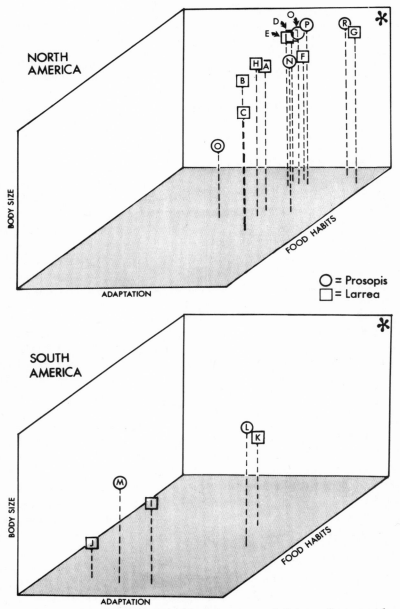

FIGURE 7-6. *Three-dimensional diagrams of the small mammal communities of North and South American* Larrea *and* Prosopis *areas. The letters refer to the communities listed in Tables 7-5 and 7-6. The three dimensions represent the ecological parameters—Body Size, Adaptation, and Food Habits. Maximum diversity would be in the upper right-hand corner indicated by the asterisk.*

mammal faunas. The arid Mojave Desert site (area Q) is more diverse, ecologically, than the Andalgalá wash site (area M), although the latter contains more species. The Andalgalá site is more diverse in the distribution and representation of body sizes of its mammal species, but these are less diverse in their food habits and adaptations. Small mammals of the ecotonal Ñacuñan area (area L) are more diverse in body size and adaptational parameters (and include twice the number of species) than those of the Mojave site, but are less diverse in food habits. All other North American sites are more diverse ecologically than either southern site.

To determine what affect the *Prosopis*-containing community has on small mammal diversity patterns, similar ecological diversity indices were calculated for creosotebush (*Larrea*) dominated communities in North and South America (Mares and Hulse, 1977). Creosote is a small shrub (< 2m high), which is known to possess a small leaf biomass and a large number of secondary chemical substances which are toxic to many herbivorous insects (Rhoades, 1977). It might be expected that if the *Prosopis* life form is important to mammals, a *Larrea* area (being almost the antithesis of a *Prosopis* habitat) would demonstrate notably different patterns with respect to mammalian inhabitants. A composite diagram of small mammal communities in North and South American *Larrea* and *Prosopis* areas is also given in Figure 7-6. Once again, the only evident differences in diversities are those noted above (i.e., Sonoran Desert small mammal communities are more diverse than their Monte counterparts). The great diversity in the semidesert New Mexico *Prosopis* sites is also apparent in *Larrea* communities. No differences in ecologic diversities are seen between *Prosopis* and *Larrea* mammals. All three niche parameters (Food, Body Size, and Adaptation) are essentially equal in both types of habitats. Analysis of Tables 7-5 and 7-6 shows that North American plant communities with *Prosopis* present are about equal to *Larrea*-dominated communities in the total number of small mammal species supported, although there is a tendency for *Prosopis* areas to support fewer species. In Argentina this pattern is reversed and *Prosopis* supports a greater small mammal diversity than *Larrea* areas (up to more than twice the total).

When species lists of the mammals of each habitat type are compared another pattern emerges. Many species are encountered in both *Larrea* and *Prosopis* habitats. Nevertheless, other groups of species are less catholic in their choice of habitats, preferring either one or the other of the two types. Generally, species which are most common to *Larrea* are adapted to extreme aridity and possess the suite of characteristics indicative of a desert specialist (e.g., independence of free water, powerful kidneys, pale pelage, bipedality, inflated tympanic bullae, and so forth; see Mares, 1973 for review). Species which tend to prefer *Prosopis* habitats, however, are more representative of areas much less xeric than a desert. These animals may be common in grasslands (*Sigmodon hispidus, Cynomys ludovicianus, Lagostomus maximus*), broad-leaved deciduous forests (*Peromyscus leucopus*), juniper woodlands (*Peromyscus truei*), or thorn scrub areas and/or rocky montane slopes (*Micro-*

cavia australis, Galea musteloides, Phyllotis griseoflavus). Logically, these species are relatively poorly adapted for extremes of either aridity or heat. Most tend to be scansorial and often climb readily into *Prosopis* and other tall desert trees as well to feed on leaves, fruit, or insects. Large trees such as these probably moderate the environmental vicissitudes characteristic of arid areas. As a food, moisture, and shade resource, the plants allow animals to enter the desert in what are essentially interdigitations of less xeric habitats within more widespread arid areas.

SUMMARY

We have accumulated evidence suggesting that *Prosopis* is an integral part of the desert ecosystem, both in North and South America. Many organisms seem to prefer *Prosopis* as a food or habitat resource when it occurs in mixed plant communities. The effect of this dominant phreatophytic tree on the system extends to both plants and animals. Because of its phenology and physiognomy, we doubt that other desert trees have as pervasive an influence upon other organisms as does mesquite. In some respects *Prosopis* seems to be more important to the integration of the desert ecosystem in the Monte than in North America. Plants and animals are associated with it in an obligate manner in that desert, while in North America this tendency is less pronounced. Certainly the effect of this species on mammalian abundance and diversity is much more marked in the southern desert. Nevertheless, in both deserts *Prosopis* serves an important function in modifying the environmental extremes characteristic of deserts and allowing species to inhabit what would otherwise very likely be an inhospitable terrain. Therefore, *Prosopis* is probably an important positive factor in helping maintain the complexity of desert ecosystems, and it is thus not surprising that today's dominant desert mammal, man, has also developed a long and close association with plants of this genus.

Chapter 8

MESQUITE IN INDIAN CULTURES OF SOUTHWESTERN NORTH AMERICA

R. S. Felger

From early pre-historic times until recent years, mesquite has served native peoples in southwestern North America as a primary resource for food, fuel, shelter, weapons, tools, fiber, medicine, and many other practical and aesthetic purposes. Every part of the plant is used. Utilization of mesquite was the common denominator among the diverse peoples of the arid southwestern lowlands (Figure 8-2), agriculturalists as well as nomadic hunters and gatherers. Because mesquite is such an important and unfailing resource, it came to figure in the everyday life of these peoples from cradle to grave. Its phenology played an important role in secular calendar events and mesquite is prominent in native oral literature.

The term mesquite or *mezquit* has become the dominant folk generic name applied to species of *Prosopis* (section *Algarobia*) in southwestern North America. Among Spanish-speaking people in Arizona, Sonora, and Chihuahua, the fruit is known as *la péchita*. This term derives from the Opata word, *péchit,* also meaning mesquite fruit (T. Hinton, personal communication; Nentvig, 1971). However, in southeastern Sonora, where more than one wild legume tree crop is harvested, cognition of *péchita* is restricted to the sweet mesocarp "pulp" of the fruit of mesquite and *guaymuchil* (*Pithecellobium dulce,* Leguminosae) which are referred to respectively as *péchita de mesquite* and *péchita de guaymuchil* (H. S. Gentry, personal communication).

While the term for mesquite varies with the different native languages, I know of no ethnobotanical linguistic differences concerning mesquite (*Prosopis* section *Algarobia*) that correspond to the different taxonomic entities (e.g., species) recognized by botanists. For example, the Cahuilla, Seri, and Quechan (=Yuma) and western Papago primarily harvested *Prosopis glandulosa* var. *torreyana* and the remaining northern Piman-speaking people and Opata used *P. velutina*. Ethnobotanically it is therefore both convenient and realistic to treat the North American members of the section *Algarobia* as a unit. In this regard, it is interesting to note that just as the various botanical taxa of section *Algarobia* are often allopatric in North America, so are the Indian groups that depended so heavily upon them (Figure 8-2).

The traditional native taxonomic concepts for mesquite in North America appear to be unambiguous: a single monotypic folk genus is generally recognized among each tribe (for a discussion of the concept of folk genus, see Berlin, 1973). Compared with modern biological species concepts, it appears that the Indians have not significantly overclassified or underclassified the mesquite. This is particularly significant because there is an axiomatic tendency in folk taxonomy to overclassify economically important species where

FIGURE 8-1. *A Papago granary basket, probably made of arroweed* (Pluchea sericea, *Compositae*). *The granary, used by the Indians at San Xavier mission near the Silver Bell study site, is elevated on stones. Baskets such as these were used to store* Prosopis *pods. For details about this granary, see Kissell* (*1916*).

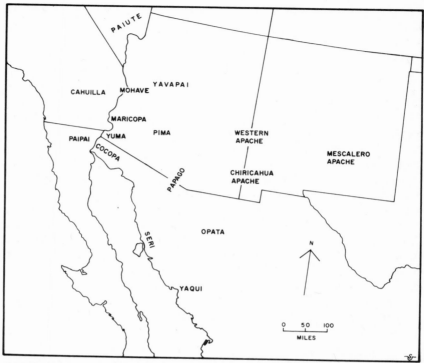

FIGURE 8-2. *A generalized map showing the area inhabited by various native peoples in the southwestern part of North America who were dependent on* Prosopis *as a food item for parts of the year.*

consistently recognizable differences can be identified (Berlin, 1973). Using the Seri and Papago or Pima as examples, the terms for mesquite, ʔáas (Seri) and *kui* (Mathiot, 1973) respectively are unanalyzable words and as such can be referred to linguistically as primary names. Such primary names characteristically suggest considerable cultural antiquity (Berlin, 1973).

Mesquite was the first plant recorded by Europeans in the American Southwest, beginning with Cabeza de Vaca's report of his epic wandering across the continent between 1528 and 1536 (Bandelier and Bandelier, 1905). Subsequent explorers and missionaries gave extensive information on mesquite, mostly as food and fuel, but also as a medicinal plant (Baegert, 1772; Bolton, 1919; Clavigero, 1789; Hammond and Ray, 1940; Manje, 1926; Nentvig, 1971; Treutlein, 1949). Much of this information on the uses of mesquite has previously been summarized (Bell and Castetter, 1937) and will not be repeated here. Rather I will emphasize here new information on the uses of *Prosopis* by the Seri and other native people in southwestern North America.

The archaeological record demonstrates that the use by, and association of, mesquite with native peoples in North America extended into ancient times (e.g., Fewkes, 1912; Harrington, 1933; Jones, 1941; Haury, 1945, 1950, 1976; Cosgrove, 1947; Bohrer, 1962, 1970, 1973a,b; MacNeish, 1964; Smith, 1967; Flannery, 1968). In recent years analysis of human coprolites has allowed reconstruction of prehistoric diets, with mesquite being a common component in such places as southwest Texas (Bryant, 1974) and Ocampo Caves, Tamaulipas, Mexico (Callen, 1969).

A detailed report of the Seri use of mesquite (Felger and Moser, 1971), gathered primarily in the native language, indicates an extensive knowledge and terminology associated with this plant. Lack of comparable detail for other groups can be attributed largely to the fact that foreign languages (i.e., English, Spanish) in which data are gathered often do not contain the necessary terminology and concepts sufficient for the native informants to express their full meaning. Furthermore, anthropologists without botanical knowledge, or botanists lacking anthropological and linguistic skills often overlook significant ethnobotanical information. However, many sources (e.g., Bean and Saubel, 1963; Mathiot, 1973; Ruth Giff, Joseph Giff, and Sally Pablo, personal communication) demonstrate that other Sonoran Desert cultures during traditional time possessed levels of terminology and knowledge of mesquite comparable to that reported for the Seri.

A caution must be given about the reported times or seasons for phenological events such as flowering and fruiting, since information was often obtained by questioning informants rather than by direct observation. Many rural people are not overly precise about the month of an event such as a particular harvest if it is many months away or past. Such information in the present report must therefore be considered preliminary. Of the vast body of knowledge regarding native use of *Prosopis* only the barest highlights are touched upon here. The principal topics that will be covered include the use of mesquite as food, for medical and cosmetic purposes, in material industries, as fuel, in recreation, and in religious and ritual practices. There is still an extensive knowledge, retained by elderly people, that remains unrecorded, and which, because of the rapid rate of acculturation, will probably not be passed down into either oral or written literature.

FOOD

Two major classes of wild edible plants may be distinguished in arid lands: the unfailing crops and the facultative crops. The unfailing class, which includes mesquite, is composed of perennial plants that can almost always be depended upon to produce large crops year after year independent of local, short-term climatic conditions such as drought or unusually cold weather (see Bean and Saubel 1961, 1972; Felger and Moser, 1971, 1976). While it might be expected that the most severe season, late spring and early summer drought, would be a time of food shortage, it was actually a time of plenty because of several dependable high-yield wild crops. Both hunter-gatherers and agricultural peoples stored surpluses from these unfailing crops. For the agricultural peoples these crops provided significant sustenance until their cultivated crops could be harvested. Most pre-contact maize-farming people also gathered wild plant foods such as mesquite (Smith, 1967).

Mesquite is the most widespread of the unfailing wild crops in the hot lowlands of southwestern North America. In many areas it was the single most important wild harvest, providing an essential and dependable food resource for the diverse peoples living throughout its range (Figure 8-2). Presumably, the very deep roots of the trees and its partial phreatophytic habit (Chapter 1) contribute to the predictable ripening of fruits shortly before or at the onset of the summer monsoon. Other nutritionally important unfailing crops in the Sonoran Desert, also harvested during the late spring and early summer dry season, include eelgrass (*Zostera marina*, Zosteraceae) among the Seri (Felger and Moser, 1973), Palmer's saltgrass (*Distichlis palmeri*, Gramineae) among the Cocopa (Castetter and Bell, 1951; Felger, 1975), and various columnar cacti including saguaro (*Carnegiea gigantea*, Cactaceae) and cardón (*Pachycereus pringlei*, Cactaceae) (Castetter and Bell, 1937; Felger and Moser, 1974a).

Facultative wild crops are plants dependent on short-term conditions such as rainfall in order to bloom and fruit, e.g., ephemerals such as *Amaranthus palmeri* (Amaranthaceae), *Plantago insularis* (Plantaginaceae), *Salvia columbariae* (Labiatae), and various perennials such as *Lycium* spp. (Solanaceae). In desert regions where climatic conditions are generally unpredictable, facultative wild crops, taken over a span of years, are less important to native people for survival than are the unfailing species.

Harvesting mesquite pods or other predictable crops during the spring dry season often meant resettlement or temporary encampment close to the resource and group cooperation (Bean and Saubel, 1972). These crops tend to produce massive quantities of fruit during a relatively short time which meant that the crop needed to be harvested quickly, and all available labor was therefore recruited. Although most wild plant food collecting was woman's work, entire families including men assisted in these arduous or complicated harvests. For many hunting-gathering societies with minimal political structure, as well as certain agricultural peoples, it was a time of coming together. Among the Quechan Indians, as the mesquite pods ripened, outlying districts were notified by runners and many people converged on the mesquite groves. The evenings were spent ". . . singing, dancing, playing games and making love" (Trippel, 1889:6-7).

The production of mesquite pods generally seems to have exceeded the desire or ability of native peoples to harvest, process, and store them. While actual quantitative data of fruit production and utilization have not been reported, fruit production does not seem to have been a factor limiting native populations. There are, however, some indications of occasional failure of the mesquite crop (Palmer, 1878). For example, it has been claimed that when the saguaro (*Carnegiea gigantea*) and mesquite crops failed, the Pima Indians of southern Arizona made long journeys into hostile Apache territory to secure food (Russell, 1908). This lack of sufficient fruiting was a rare event and might have been due to severe winter freezing. During 1975 and 1976 extensive stands of mesquite and foothill palo verde (*Cercidum microphyllum,* Leguminosae) near Tucson failed to produce fruit. This seems to be attributable to unusually cold weather in winter and spring. However, in certain protected sites and at lower elevations productivity was high and it was possible to harvest substantial quantities of fruit.

In the subtropical thorn scrub regions to the south of the Sonoran Desert, such as in Sinaloa, Mexico, the traditional use of, and dependence on, mesquite as food diminished because of the abundance of other more desirable or preferred legume tree crops such as *quaymuchil* (*Pithecellobium dulce*) (see Gentry, 1942, 1963; Felger, 1976). Tropical and subtropical crops and the more dependable monsoon-supported agriculture made reliance on mesquite less of a necessity. Yet, in the hot subtropical lowlands of western Mexico south of the Sonoran Desert, such as southwestern Sonora and western Sinaloa, historical accounts as early as the sixteenth century document extensive utilization of mesquite as a human food resource (Winship, 1896).

At higher elevations, and eastward and northward into temperate regions, desert scrub gives way to grasslands. Here, the collecting of wild grass seed, and the hunting of animals feeding on these grasses, more or less substituted for mesquite dependence farther south. Towards the Pacific coast the distribution of mesquite halts with the waning of the summer monsoon; and in these Mediterranean climates, as well as locally along the eastern margin of the Sonoran Desert, acorns (*Quercus* spp., Fagaceae) generally assumed the role of the mesquite pods (Bean and Saubel, 1972). At its geographic limits, mesquite also served as an item of commerce. For example, the Wanikik Cahuilla of southern California ". . . who had some mesquite in their area, regularly traded their abundant acorn and piñon crops for mesquite from their neighbors in the desert" (Bean and Saubel, 1963:56).

Individual mesquite trees, as well as certain geographically-defined populations, were discovered and identified as having superior flavor and yield. These special trees and populations were specifically sought year after year, yet there is no indication of any attempt at selection or cultivation. The Seri Indians know that the mesquite groves on the coastal plains between Kino Bay and Tastiota produce sweeter, better tasting fruit than those of other populations, and certain groups of the Seri made special harvest encampments there (Felger and Moser, in prep.). At harvest time the Maricopa Indians of southern Arizona sought certain trees known to produce exceptionally large or sweet pods (Castetter and Bell, 1951). The Cahuilla knew that desert mesquite groves produced greater quantities and "tastier" fruit than trees occurring at higher elevations (Bean and Saubel, 1963), which would be at or near the desert-chaparral ecotone. Many Southwestern people, such as the Cahuilla (Bean and Saubel, 1963) and Pima (Sally Pablo, personal communication) do not consider every tree of sufficient quality to merit harvesting. Among the Pima, fruit was gathered only from trees with the largest and thickest pods, and pods with reddish streaking are considered best of all (Sally Pablo, personal communication).

The native mesquite of Baja California (at least in the Central Gulf Coast) apparently produce bitter pods which were seldom eaten by the Indians (Baegert, 1772) and plants with sweet-tasting pods from the Mexican mainland were introduced in Spanish colonial times (Aschmann, 1959). After the Spanish conquest, mesquite was brought from the Yaqui region, or some other part of the mainland coast, and planted at the mission at Loreto and at one or two other Baja California missions (León-Portilla, 1973). These trees yielded fruit of good flavor which were subsequently eaten by the local Indians. The concept of sweet-fruited mesquite from the mainland and bitter ones from Baja California still persists.

Rural people in north-central Sonora, specifically at Cinoquipe on the upper Rio Sonora and Cucurpe on the upper Rio San Miguel, distinguish between mesquite trees yielding bitter-tasting pods and those with sweet, edible pods. Trees with sweet pods are said to be far more common than those with bitter pods, and are distinguishable only by taste.

There is considerable ethnological variation in details regarding the collection, preparation, and storage methods of the pods and their products. The following brief description of traditional Seri Indian harvest and preparation techniques (Felger and Moser, 1971), will serve to indicate generalized methodology. For the sake of simplicity the Seri terminology is omitted. However, it should be noted that Seri mesquite-associated terminology is far more extensive than it is in English or Spanish (Felger and Moser, 1971). The Seri are a Hokan-speaking people living along the east side of the Gulf of California. Traditionally they are a semi-nomadic, seafaring, hunting and gathering people (McGee, 1898, 1971; Kroeber, 1931; Griffen, 1959; Spicer, 1962; Moser, 1963, 1976; Bowen, 1976). Today mesquite is seldom harvested—it is hard work and comes at a time of very hot weather.

An indication of the former significance of the mesquite pod in Seri culture is evidenced by the fact that they have names for eight stages of growth of the fruit from the youngest (less than 3 cm long) to the fallen, rotting pod. In the second youngest stage, the pods are tied into small bundles and cooked with meat. Full-sized green pods are mashed in a mortar formed in bedrock or hard earth. The pounding is accomplished with a pestle about 1 m long made of mesquite or ironwood (*Olneya tesota*, Leguminosae). After the green pods are mashed, they are cooked in clay pots. Ripe pods picked from the tree are prepared in the same manner.

The most commonly utilized form of the fruit is the dry fallen pods which apparently have the highest mesocarp carbohydrate content. They are gathered in large quantity in shallow tray-baskets (Moser, 1973). The load is built up by placing sticks—in this case probably mesquite sticks—ver-

FIGURE 8-3. *Papago Indian Juanita Ahill uses a stone pestle in a bedrock mortar grinding mesquite pods into flour (1975).*

tically into the load of pods around the edge of the basket, then piling on more pods, and, in turn, holding these in place with additional vertically placed sticks. There are numerous variations on the methods of preparation, the most common of which are given below.

The pods may be pounded in mortars with wooden pestles and then chewed, the sweet juice swallowed, and the pulp discarded. However, the usual method is to toast the pods before pounding. To toast the fruits, the

Indians clear the ground, light a fire in the center, and then remove the coals. The dry pods are then piled on the hot earth. At the same time, four piles of sand are placed around the area and fires are burned on each pile to heat the sand. The pods are then sprinkled with the hot sand, allowing the heat from the fires plus the hot sand to toast them. Toasting probably helps eliminate potential loss from bruchid beetles (Chapter 6), but there appears to be no knowledge among the Seri that the toasting is for such purpose. Toasting the pods greatly facilitates grinding. The Seri moon or month known as "the moon to sprinkle [sand] " derives its name from the sprinkling of hot sand on the mesquite pods. This time of year, approximately the month of July, is the beginning of the new year for the Seri and other Sonoran Desert peoples such as the Pimans. It is significant that the year begins when the common desert trees and shrubs drop their seeds and the brief Sonoran Desert monsoon begins—for it is the time of greatest renewal of life in this part of the world.

After the pods are toasted, they are carried to a bedrock mortar to be pounded with a wooden pestle of mesquite or ironwood (*Olneya tesota*). A large pile of pods is placed in the mortar and more are placed on the ground surrounding the hole (Figure 8-3). Several women may then pound at the same time, each at her mortar. As the pods are crushed, more are added from those piled around the mortar hole. After the pods are mashed they are placed between deer skins to prevent spoiling in the hot and often moist July wind blowing off the Gulf of California. The pounding continues until all the pods are mashed. The woman then places the pestle across the mortar hole. Mashed pods or pulp are put in a basket and winnowed by gently tapping the basket against the pestle. Flour from the mesocarp (Figure 6-2) falls into the mortar hole, and the seeds enclosed in the stony endocarp plus pieces of fiber and exocarp remaining in the basket are set aside on a deer skin. The flour is again winnowed until it is nearly pure and then placed in a pottery vessel in which it may be stored for some time, retaining its characteristic aroma and taste. One man estimated that two women, working with a man who keeps them supplied with mesquite pods, are able to prepare about 40 kg of mesquite flour in a day.

The Seri place the flour in a large shallow basket, mix it with water, and knead it into a dough which is shaped into rolls often about 20 cm long and 5 to 10 cm thick, or into round cakes. The rolls and cakes are dried immediately in the sun so that they will not spoil. When dry they too can be stored in pottery vessels for an extended period. In former times families often had two or more large vessels filled with mesquite cakes hidden in caves for times of need. These narrow-mouthed storage vessels had lids fashioned from a large clam shell, flat stone, or piece of pottery. The lid was tightly sealed with lac from a scale insect, *Tachardiella larreae*, which is often found encrusted on the stems of creosotebush (*Larrea tridentata*). The lac is plastic when heated but hardens again on cooling, forming a strong bond closely akin to commercial sealing wax (Euler and Jones, 1956; also see Bohrer, 1962, and Standley, 1923). The lid is easily removed by severing the hard, dry lac with a heated knife blade (Felger and Moser, in prep.). The Seri and other Indians also obtained lac from the stems of *Coursetia glandulosa* (Leguminosae), produced by the scale insect *Tachardiella fulgens* (Felger and Moser, in prep.; Euler and Jones, 1956). A third kind of lac was obtained from the stems of arrow weed (*Pluchea sericea*, Compositae) (see Euler and Jones, 1956). Preservation of a wide variety of dry plant-derived foods, both cooked and uncooked, in sealed pottery vessels was widely practiced in southwestern North America (Euler and Jones, 1956). Such vessels have generally been recovered from rock shelters.

After separation by winnowing, a second pounding, this time of the stony endocarp, breaks the pods open and frees the seed. The seed is separated from the endocarp by another winnowing and then ground on a grinding stone (*metate*). Flour resulting from grinding the seeds is mixed with water, some dry mesocarp flour, and the mixture drunk. While considerable effort must be expended to procure a significant quantity of seeds, they are nutritionally very rich, with a protein content of about 40 percent (Walton, 1923; Earle and Jones, 1962; Jones and Earle, 1966; Felger, 1975). However, utilization of the actual seed was not as widespread or frequent as use of the mesocarp tissue. Verification of use of the seed is often difficult to determine from the literature because of indiscriminate usage of the word "bean" to mean either the entire pod, the endocarp and the enclosed seed, or perhaps the actual seed itself. Among the more reliable reports of utilization of the mesquite seed are Bell and Castetter (1937:22-23), Bohrer (1970), and Felger and Moser (1971).

Extensive utilization of the actual seed seems to have been achieved by the Amargosan-Pinacateño people who occupied the 1000 km² Pinacate lava fields in extreme northwestern Sonora (Hayden, 1967, 1969). These people developed "a unique grinding tool, termed a 'gyratory crusher' . . . . [which] resembles a perforated mortar, either in slab or block form, in which a wooden pestle with a projection extending through the perforation in the mortar base was gyrated, the projection providing leverage against the under rim of the hole, to grind mesquite pods. The crusher underwent modifications, and its use seems to have been discontinued with the disappearance of mesquite forests at the end of the Yuman I period, about A.D. 11-1200" (Hayden, 1967:154). Use of the Pinacate gyratory crushers (Figures 8-4 and 8-5) probably spanned three to four millenia. Julian Hayden and I recently processed oven-heated mesquite pods in one of the prehistoric Pinacate crushers. The pods were broken into two or three pieces, fed into the crusher, and pulverized with a wooden pestle (see Figure 8-5). We were able to process several handfuls of pods with very little effort in less than a minute. The stony endocarps were readily broken open, yielding seeds in considerable quantity. The mesocarp flour and seeds could then easily be separated by winnowing. Similar stone implements have been found elsewhere in Sonora, as well as in the Old World, such as the Khazinah phase deposits in Iran, dated six to seven millenia before present (Hole, et al., 1968; Hayden, 1969).

FIGURE 8–4. *Prehistoric gyratory crushers* in situ *at Tinaja de Tule, Sierra Pinacate, Sonora, Mexico. These stone crushers were used to grind* Prosopis *pods in the Pinacate Regions until about 1200* A.D.

Gyratory crushers may be more widespread, and the suggestion that Old World implements may have been used to process pods and seeds of *Prosopis farcta* or other legume tree crops including the carob tree (*Ceratonia siliqua*) is most intriguing (Hayden, 1969).

The fibrous material obtained by the Seri from the first pounding of the pods is chewed for the sweet flavor and then discarded. This sweetness results from adhering bits of mashed mesocarp. Or these fibers may be pounded a second time, water added, the fibers sucked, and the pulp then discarded. Sometimes this chewed fibrous pulp is saved, mixed with sugar, toasted, and then added to water to make a special drink. (Since neither sugar nor honey were available in earlier traditional times, sugaring is obviously a recent modification.) During pre-contact times mesocarp tissue provided these desert people with one of their few sweets.

In addition to numerous variations on the methods outlined above, various refreshing drinks were prepared from the pods. After the first pounding of the pods, the Seri sometimes placed endocarps with seeds into a pottery vessel with water, weighted them down with a stone, and left them to stand until the water became sweet. This juice was a special treat to the children because at that time of year they had little or no access to sweets. Sometimes men would prepare this beverage and allow it to further ferment for several days before drinking it. This, and other mildly fermented, beer-like drinks prepared from the pods were of widespread use among southwestern Indians

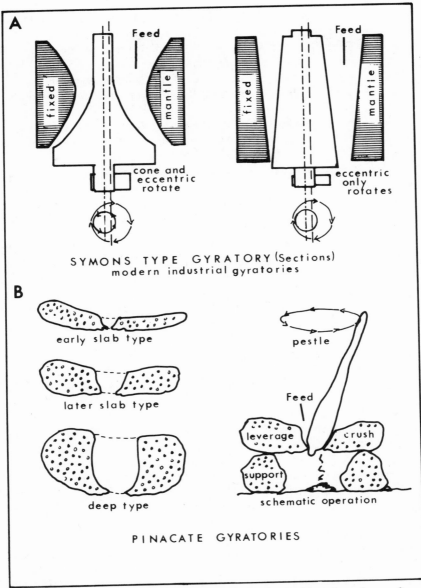

FIGURE 8-5. *A diagramatic representation of a prehistoric and a modern gyratory crusher showing the similarities in the principal of operation.*

(Havard, 1896; Castetter and Opler, 1936; Bell and Castetter, 1937). In general, it does not appear to have been intoxicating and was probably very low in alcoholic content. Most of the reports (see above) refer to drinks prepared from mesquite as cool and refreshing. "Since the pods were preserved [dried], this drink was part of the year-round diet of the Cahuilla" (Bean and Saubel, 1963:58), and was drunk continuously ". . . during the hot summer months" (Barrows, 1967:73).

Occasional references to highly intoxicating drinks prepared from mesquite pods or flour by primitive methods of fermentation should be verified. During colonial times or later, mesquite flour was sometimes used as a sugar substitute in preparing grain-derived liquors. Earlier in this century some Seri men collected mesquite pods for a group of Mexican men with a portable, and presumably illegal, still. The distillate was a highly intoxicating liquor, but ensuing drunkenness and disagreement soon terminated the venture (Felger and Moser, in prep.).

There are numerous reports of "bread" or cakes made from mesquite meal or flour which were often like ordinary loaves of bread (Bell and Castetter, 1937). Presumably the flour was usually prepared from the mesocarp of the ripe, fallen pods. The texture varied according to the particular method of flour preparation, namely the coarseness of grinding. Mesquite cake is usually somewhat yellowish or gold-colored and farinaceous with a bran-like texture. The taste is sweeter than that of ordinary wheat bread or tortillas made from maize. Mesquite cake requires no baking: the water and flour mixture is simply allowed to dry.

Both the pods and prepared mesquite flour and cake were stored for future use. Storage time was a year or more (Felger and Moser, 1971; Bean and Saubel, 1963), but generally there was no need to store the pods or their products beyond the time of the next harvest (Ruth Giff, personal communication). Settled people such as the Cahuilla, Cocopa, Pima-Papago, and Quechan stored the whole pods in large wicker or basketry granaries atop their homes (Figure 8-6) or elvated off the ground next to the house (Figure 8-1), (Bartlett, 1854; Bean and Saubel, 1972; Castetter and Bell, 1951; Forde, 1931). Screwbean (*Prosopis pubescens*), maize, and wheat were also stored in these granaries (Ruth Giff, personal communication). Many settled people such as the Maricopa and Pima used the roofs of their houses and granaries to dry the pods, but if rain threatened, the pods were taken inside to prevent spoilage (Bartlett, 1854; Castetter and Bell, 1951). Early settlers in the Moapa Valley, Utah, remarked about seeing enormous conical mesquite cakes, weighing from fifty to sixty pounds apiece. These dried cakes were stored in grass-lined pits in rock shelters along the rear wall of Paiute wikiups (Stuart, 1943).

Damage by bruchid beetles (see Chapter 6) to stored pods and their products is well documented in the literature (e.g. Castetter and Bell, 1937; Bean and Saubel, 1972). A number of nineteenth and early twentieth century

FIGURE 8-6. *A Papago wattle-and-daub construction house at San Xavier near Tucson, Arizona. The posts and larger poles are undoubtedly mesquite wood. The large basketry granaries were commonplace features of traditional houses in this region and were used to store* Prosopis *pods (1894).*

observers reported that the presence of substantial quantities of the larvae made little difference to the Indians, and were commonly ". . . accepted as an agreeable ingredient" (Castetter and Bell, 1937). More recently, because of changing prejudices, bruchid-containing food has become unacceptable to such people as the Pima (Ruth Giff and Sally Pablo, personal communication) and the Cahuilla (Bean and Saubel, 1972). However, the Seri custom of heat-treating the pods (see above) seems to have effectively controlled the ever present bruchids. Numerous bruchid holes are clearly evident in the carbonized remains of mesquite pods from various archaeological sites such as those recovered from the thirteenth century Point of Pines Ruin in northern Arizona (Bohrer, 1973).

Mesocarp-derived flour provided considerable caloric value (see Chapter 6) although the protein content does not seem to be nutritionally significant (Walton, 1923; Felger, 1975). Protein intake was largely obtained from wild animals (including fish) and seeds of other desert plants and crops, and in the case of the Seri and some other groups also from mesquite seeds. A convenient and abundant calorie-rich component, such as provided by the mesocarp-derived flour, would be highly desirable and a significant factor for subsistence in these very arid environments, particularly during the dry season. The Seri say that food from the mesquite makes children fat and their skin light in color, both desirable conditions to the Seri (Felger and Moser, 1971).

In the hottest regions of the Sonoran Desert and where winter is particularly mild, such as in Cocopa territory at the delta of the Colorado River, the mesquite harvest began in mid-June. Further north in southern Arizona in the Mohave, Quechan, and Pima regions, the harvest usually commenced in late June and continued through most of July. In the Seri region in western Sonora, a second, but lesser, crop of pods is generally produced at the end of summer, usually in early September. Pods were harvested from this crop too (Felger and Moser, 1971). Among the Cahuilla, in southeastern California, "the mesquite is available for gathering during three months of the year, June through August, depending on the ripening times of different areas. In the lower Colorado Desert [ca 60m below sea level to about a hundred meters above sea level], it is ready in June; at Palm Springs [130 to 160 m elevation], in July; and near Whitewater [ca. 400–500 m elevation], about August" (Bean and Saubel, 1963:57).

The most common method of harvest consisted of gathering the dry pods shortly after they had fallen to the ground (Felger and Moser, 1971; Bell and Castetter, 1937). However, the Yuman people also used wooden hooks to pull down fruit-laden branches in order to pick the pods (Castetter and Bell, 1951). The Cahuilla often picked green pods directly from the tree, and then returned several weeks later to harvest the dry, fallen pods (Bean and Saubel, 1963). When picked green, the pods were ripened artificially by placing them in the sun.

The time of mesquite harvest was often extended by robbing packrat (*Neotoma* sp.) nests (Castetter and Bell, 1951; Felger and Moser, 1976, see Chapter 7). The nest was pulled apart and the packrat's supply of mesquite pods and other edible fruit or seeds was collected. In earlier times the hapless packrats were also added to the larder. The Seri robbed packrat nests up to several months after the mesquite harvest, usually in late fall. At least in the Seri region, the packrats conveniently segregate their caches according to species in easily removable piles (Felger and Moser, 1975).

Mesquite flowers were also eaten by certain people. The Cahuilla roasted the flowers in a pit with heated stones, then pressed them into balls which could be stored in pottery vessels, and at a later date eaten after being boiled in water (Bean and Saubel, 1972). The Pima ate the flowrs by stripping them from the inflorescences between the teeth (Russell, 1908), and also sucked them "because they are sweet" (Curtin, 1949:95). The Pima are also reported to have used the inner bark as a substitute for rennet (Russell, 1908). Seri children like to chew a kind of light-colored mesquite gum which is said to be bitter, spit it out, and then drink water. They say the water then tastes sweet (Felger and Moser, 1971). The Pima also made a candy from a white mesquite gum (Russell, 1908; Bell and Castetter, 1937). It was ". . . consumed raw or

. . . prepared by covering with hot ashes, causing the gum to swell" (Curtin, 1949:95). The Kiowa (Vestal and Schultes, 1939) and others valued the gum for chewing.

Among agricultural people the harvest of mesquite was largely replaced by that of winter wheat (William Doelle, personal communication), and for both farmers and hunter-gatherers, the trade and wage economy brought about shifts to store-bought or bartered foods. Mesquite was not among these foods. However, subsequently during times of hardship both Indian and non-Indian people have periodically relied on mesquite, and as such it appears to be the most important and frequently harvested wild food plant resource. For example, in the nineteenth century non-Indian people along the San Pedro Valley in Arizona relied on mesquite pods during times of Apache raids (Alexander Russell, personal communication). During the depression of the 1930s people at Ajo, Arizona, some of them Papago Indians, likewise made use of mesquite pods (Thomas Hinton, personal communication).

The screwbean or *tornillo* (*Prosopis pubescens*) was also an important item in the native diet of portions of southwestern North America (Bell and Castetter, 1937; Castetter and Bell, 1951; Bean and Saubel, 1972). However, its range is not as extensive as that of the mesquite, nor is it usually as abundant (Hicks, 1961). Screwbean pods ripen in mid- to late summer. They are prepared in much the same manner as are mesquite pods and are reported to be sweeter than those of mesquite. Although it was locally an important wild food, it was generally not as significant as mesquite. Since screwbeans grow in lowland places subject to frequent flooding, it is generally absent from the territories of the nomadic hunter-gatherer peoples of the Sonoran Desert such as the Seri and the western Piman-speaking peoples.

MEDICINE

Mesquite has been extensively used by American Indians for a wide range of medicinal purposes (Hrdlička, 1908; Standley, 1923; Bell and Castetter, 1937; Vogel, 1970). Mesquite gum, the black pitch or flux, herbage, roots, and bark have all been employed, usually in an aqueous solution or tea. With minor exceptions, neither flowers nor fruits were used medicinally. Dry pods were boiled and the decoction used as a "bleach" after severe sunburn by the Pima Indians of southern Arizona (Curtin, 1949:94).

Preparations involving mesquite, as well as most other native medicines in the Sonoran Desert, are usually simple. Absence of complicated formulae facilitates non-professional and individual preparations and administering, and the rapid diffusion of information through a culture (Felger and Moser, 1974b). Among Sonoran Desert people such personal medicines generally do not involve a shaman (medicine man).

While traditional remedies, such as those involving mesquite, are less practiced today because of acculturation, they may still be resorted to, at least on occasion. In many cases traditional medical practices have continued even after other vestiges of the original Indian culture have disappeared. Selected, common, traditional medical uses of mesquite, most of which may be presumed to be ancient, are briefly mentioned below.

The most common medicinal use of mesquite leaves and gum (occasionally mixed with non-vegetal substances such as certain minerals) is for eye ailments. The Aztecs used such a mixture when the eyes were hot and painful from sickness (Emmart, 1940; Anderson and Dibble, 1950-1959). The Mescalero Apache and people in Baja California ground the leaves into powder, wrapped the powder in a thin cloth, added water, and squeezed the liquid into an afflicted eye (Hrdlička, 1908; Clavigero, 1789), whereas the Paipai in northern Baja California and the Pima in Arizona boiled the leaves in water and washed an inflamed eye with the solution (Owen, 1963; Castetter and Bell, 1937). Similar usage persists in the folk medicine of Mexico (Martinez, 1959; Standley, 1923). Certain Indians and Spanish Americans in New Mexico mashed tips of the branches (probably with leaves) in water, allowed the liquid to stand overnight, and applied the solution to sore eyes (Bourke, 1894).

The most common application of the gum for sore eyes was to dissolve it in water and apply the solution as eye drops, eye wash, or lotion (Hrdlička, 1908; Standley, 1923; Balls, 1962). The Pima sometimes placed a small piece of gum in the corner of an afflicted eye and kept it there as long as the patient could bear it. Tears dissolved part of the gum, coloring the eye brownish (Hrdlička, 1908). The Papago used the gum for soreness of the eyelids; the Maricopa applied dried powdered juice of the gum to the eyelids but not the eye itself (Hrdlička, 1908). The Seri use eye drops made from a certain type of mesquite gum described as a milky sap which has hardened like resin (Felger and Moser, 1971, 1974b).

The use of emetics and purgatives to "cleanse the system" was widespread in traditional Indian medicine (Vogel, 1970). Mesquite leaves, but more often the bark and gum, were used for this purpose. The Seri made a tea from mesquite leaves which was drunk as an emetic (Felger and Moser, 1974b). The Pima used a decoction of the inner bark as both an emetic and cathartic (Russell, 1908; Standley, 1923); and the black gum, boiled in water, was imbibed to purge the system (Curtin, 1949:94). The Paipai made similar use of bark decoctions (Owen, 1963). At least certain Yuman people drank an infusion of mesquite leaves to relieve painful micturition (Forde, 1931).

Mesquite gum dissolved in water has also been widely used as a remedy for sore throat (Martinez, 1959; Balls, 1962). For this purpose the Pima drank hot tea prepared from the sap (Hrdlička, 1908), used it as a gargle

(Standley, 1923), and also drank it as a remedy for respiratory afflictions (Curtin, 1965).

During the first four days following birth, a Cocopa baby was given a decoction prepared from the inner boiled bark. A woman would dip her finger in the liquid and let the baby suck the finger (Gifford, 1933). The Pima placed powdered mesquite bark mixed with sand or other substances on a newborn baby's umbilical cord and later the navel to prevent soreness (Hrdlička, 1908). Mesquite gum was similarly used to prevent infection (Curtin, 1949). Umbilical hernias, a rare disorder among the Maricopa children, were treated by administering tea prepared by boiling mesquite roots cut into small pieces (Hrdlička, 1908). The Pima used a solution of the gum as a disinfectant for open wounds (Standley, 1923).

Diarrhea and stomach disorders, often chronic, continue to be widespread among many Indians and economically disadvantaged Mexican people. Mesquite leaves, gum, and bark have often been used for such problems. The gum, dissolved in water, is used for diarrhea in Mexican folk medicine (Martinez, 1959). In the eighteenth century the Opata in Sonora prepared a "froth," probably from the gum, as a remedy for ulcers (Nentvig, 1971). The Comanche (a tribe living farther north of the area included in Figure 8-2) used preparations of the leaves to neutralize stomach acidity (Carlson and Jones, 1940).

As a remedy for diarrhea the Pima drank a tea prepared from young mesquite roots (Curtin, 1949) and an infusion made from the gum (Standley, 1923). They also prepared a cooling drink from the crushed leaves for stomach trouble (Curtin, 1949). The Papago made use of the "white" inner bark by pounding it as fine as possible, boiling it with salt, and taking a dose daily before breakfast to ward off chronic indigestion (Hrdlička, 1908). The Papago also drank a decoction of the powdered white inner bark as an internal antispasmodic (Curtin, 1965).

As a laxative, the Seri drank the liquid produced by the bark of green or young branches which is cut into long strips, tied into rolls, and soaked in water (Felger and Moser, 1974b).

The sap or gum was widely applied topically for skin disorders, apparently with a considerable degree of success. The boiled sap (=black pitch or sap) was applied directly to pemphigus and other sores of children (Hrdlička, 1908); the black gum, boiled in a little water, was applied to sore lips, chapped and cracked fingers, and as a lotion for "bad disease" (Curtin, 1949). The Aztecs in Mexico used a decoction of the leaves to restrain excessive menses, and the bruised bark as an astringent (Curtin, 1949).

Mesquite-derived medicines were apparently rarely used for the epidemic diseases of Old World origin which have devastated American Indian populations, although the Paipai drank tea prepared from the boiled bark as a remedy for smallpox and measles (Owen, 1963).

COSMETIC

Many Indians were particularly concerned with gray or sun-bleached hair, and the use of mesquite gum hair plaster by both sexes was widespread (Palmer, 1878; Bourke, 1889; Hrdlička, 1906; Russell, 1908; Densmore, 1932; Gifford, 1932; Spier, 1933; Bell and Castetter, 1937). The plaster was usually prepared by boiling black mesquite gum or pitch, or bark covered with the black pitch, and river mud—preferably black mud. This plaster was used to kill lice, cleanse the hair, make it glossy, and dye it black. The Yavapai believed it necessary to wash out the plaster and put on a new application before daylight lest the sun "burn up the hair and turn it red or yellow" (Gifford, 1932:229). The Yavapai practiced abstinence while the plaster was on the hair. They used three or four consecutive applications (Gifford, 1932). The Quechan prepared the plaster from boiled mesquite gum and mistletoe (*Phoradendron californicum,* Viscaceae), (Densmore, 1932). Apparently the essential ingredient is the black mesquite gum or pitch.

Mesquite charcoal and sometimes the leaves were used in traditional tattooing practices. Southern California Indians used mesquite leaves to produce blue-colored facial tattoos (Palmer, 1878). The skin was punctured with a cactus spine and then moistened leaves were rubbed over the area to give the desired blue color. Mesquite charcoal was widely used for tattooing, mostly for girls during their puberty rites (Russell, 1908; Gifford, 1931, 1933; Forde, 1931), but various other kinds of charcoal such as willow (*Salix* sp., Salicaceae) probably could also have been used (Russell, 1908). The designs commonly consisted of thin lines on the chin or a vertical line on the chest. Mesquite spines were sometimes used as tattoo needles (Gifford, 1931).

The Seri made black facepaint from pieces of mesquite bark placed in water with chunks of mesquite pitch (the partially hardened oozing black exudate) and slowly cooked. Sugar was added to help darken the mixture. When sufficiently thick, the black mass was allowed to dry for a day or two and then formed into cakes or patties which could be stored until needed. When one of these cakes was rubbed into a bit of water on a stone, a paste resulted which was used as the facepaint. Facepainting among the Seri was an intricately developed art form and was practiced by both men and women until the mid-twentieth century (McGee, 1898, 1971).

RECREATION

Wooden balls, commonly made of mesquite wood or gum, were widely used in various native sports or games. Relay races in which men kicked the ball along a prescribed course were common. The Cocopa shaped a mesquite-wood ball, about 7 to 8 cm in diameter on a grinding stone (*metate*). It was sometimes inlaid with shell beads imbedded in arroweed (*Pluchea sericea*) gum (Gifford, 1933). The Papago made a similar game ball from either mesquite wood or a pebble imbedded in mesquite gum (Bell and Castetter, 1937). Seri women had a race, similar to the men's relay race (above), in which they rolled hoops made of strips of mesquite root. The hoop was rolled with a slender stick (Felger and Moser, 1971).

The Pima had a gambling or guessing game in which a small ball of black mesquite gum was hidden in one of four reed tubes. The object of the game was to guess which tube contained the gum ball (Culin, 1907; Russell, 1908). Maricopa men made dice from mesquite root wood, each one being about ". . . 7 inches long bearing marks burned into the face . . ." (Spier, 1933:342).

Papago girls made dolls of mesquite leaves tied with strips of corn husks for arms, legs, and heads (Underhill, 1939). Seri boys use a long mesquite spine, attached to a long stick or reedgrass culm (*Phragmites australis*, Graminae), as a toy harpoon for capturing small fish and crabs.

Certainly one of the most significant and pleasant aspects of mesquite in the lowlands of southwestern North America is the welcome shade and shelter it provides from the sun during the long hot season of the pre-monsoon drought when few other trees have foliage. In the Seri calendar, the moon or month corresponding approximately to May can be translated as the "moon to sit under shelter," implying that it is the month to sit in the shade of a ramada, such as one made of mesquite, and enjoy the pleasant spring weather.

MATERIAL INDUSTRIES

Black mesquite pitch and gum have been widely used to decorate pottery, and this is particularly well-documented among the Piman-speaking people and some of their neighbors (Hrdlička, 1906; Russell, 1908; Densmore, 1932; Castetter and Underhill, 1935; Fontana et al., 1962). The detailed report on Papago pottery (Fontana et al., 1962:77–78) describes the process of decorating pottery with mesquite-derived paint as follows, and may be taken as a model:

. . . Potters used to prepare red paint by mixing red hematite . . . with mesquite gum in the same way black paint is prepared with black mesquite bark.

Black paint, . . . is made with mesquite bark that has blackened on the tree where sap has oozed out, and with mesquite gum.

The potter strips the black bark from the mesquite trees with the help of a knife, and takes it, along with balls of clear golden mesquite gum picked from the branches, back to her house. She boils water in a small can, perhaps 1/2 to 1-gallon size, adding five or six strips of the black mesquite bark while the water is heating. She allows the water with the bark in it to come to a full boil, then removes it from the heat. At this juncture she adds a small handful of mesquite gum, three or four balls, to the brew. She puts the can back on the heat and brings the paint to a boil from three to five more times.

When the paint nears readiness, the potter tests the color and consistency by dipping a stick into the can of paint and making a streak on a pot kept handy for this purpose. If the paint is not too runny, and if it has what seems to the potter to be the right color, she removes the can from the heat for the last time.

Some painters use feathers for a paint brush, but most use the unaltered tip of a devil's-claw (*Martynia*). They do not chew the tip or in other ways make it more brushlike. The painting implement is dipped into the paint, and a design is applied . . .

In all cases paint is applied to fired pottery which is generally, but not always, slipped. When the painting job is done, the pot is fired a second time to fix the paint. This final firing consists only of putting the painted surface of the pot directly on hot coals or in revolving the pot over flames for three or four minutes. Not only does this turn the mesquite paint a coal black color, giving it a shiny appearance where properly fired, but it tends to alter the color of the pot slightly, darkening it.

The Papago collected mesquite gum after the summer rains and made it into balls which were stored for future use. The balls could be boiled until syrupy and used like fresh sap. Mesquite gum was also used as a paint for war shields (Castetter and Underhill, 1935).

The cakes of black pigment made by the Seri from mesquite pitch, described above (see "Cosmetic," above), have also been used as a black basketry dye. The cakes are dissolved in boiling water and strips of natural basketry splints, prepared from stems of *Jatropha cuneata* (Euphorbiaceae), are immersed in the dye pot, which results in dying them light gray. However, black is preferred to gray. To obtain black, basketry splints which have previously been dyed reddish-brown with dye prepared from the root of *Krameria grayi* (Krameriaceae) are overdyed with the mesquite pitch preparation (Felger and Moser, 1971; Moser, 1973).

The Seri made extensive use of various kinds of strong cordage fashioned from mesquite roots (Figure 8-7). The root, with bark removed, was chewed to soften the fibers and then twisted (spun) into twine or rope. The Seri have specific names for three kinds of double-strand mesquite-root cordage, and another name for a three-strand rope. This cordage was used for harpoon lines for hunting sea turtles and spearing large fish, and for binding certain harpoon points to the shaft. A certain fish line, also given a specific name, was used for stringing fish as they were harpooned. The fisherman fastened one end of the line around his waist while wading through the water spearing fish. He strung fish on the line with a creosotebush (*Larrea tridentata*) point and trailed the fish several meters behind him in the water, so that if sharks were attracted he might not be attacked (Felger and Moser, 1971, in prep.).

Mesquite cordage was also used by the Seri to lash together bundles of

FIGURE 8-7. *Fiber and twine made from mesquite roots (Seri, ca. 1965). The upper figure shows a section of root 30 cm long with the bark removed and after it has been pounded with a* mano *(grinding stone) that loosens the fibers. The lower figure is a 90 cm long piece of two-stranded twine.*

reedgrass (*Phragmites australis*) for seagoing reed boats or *balsas*. The Seri balsa was sometimes 10 meters in length. The bundles were made by inter-weaving individual reeds, tying them with mesquite cordage, and then lashing three large bundles together to form the boat. Seri men also made temporary rafts for hunting sea turtles. These were made from pieces of driftwood tied together with mesquite cordage. After use the hunter would dismantle the raft in order to save the twine, which was regarded as a valuable possession (Felger and Moser, 1971, in prep.; also see McGee, 1898, 1971).

Seri use of mesquite twine also included waist cords and carrying nets. These nets were used to suspend the large, exceedingly thin, traditional Seri vessels, known as eggshell ware, from carrying yokes. This pottery was made as thin as possible, presumably to reduce weight, and was traditional in every-day use for carrying water (Bowen and Moser, 1968). On occasion, a small child, placed in a shallow tray basket, was carried across the desert in one of these nets swinging from a carrying yoke, the other end balanced with a water-filled pottery vessel or cargo (Moser, 1970). The assertion of Seri use of fabric woven from mesquite fiber (McGee, 1898, 1971) is not at all sub-stantiated by present data (Felger and Moser, in prep.). The Kamia made strong twine from the inner bark which was first soaked in water for a month (Gifford, 1931). Spinning was done on the bare thigh. Mesquite roots and bark were occasionally used in basketmaking, such as among the Cocopa (Chittenden, 1901) and the Mohave (Merrill, 1923).

Immediately following the birth of the child, a Quechan, Pima, or Mohave father made a cradle from mesquite wood by bending a slender, flexible mes-quite branch into an elongated U-shape and lashing to it flat slats of mesquite wood forming a flat bed. Hoops of mesquite were curved over the upper end as an awning (Forde, 1931; Bell and Castetter, 1937). Among the Yavapai the cradle, made by women, had a frame of mesquite wood. Many other groups variously used mesquite wood, including the flexible root, in cradle construction (e.g., Gifford, 1931; Kissell, 1916; Felger and Moser, 1971). In prehistoric times the Hohokam likewise placed their babies on mesquite cradleboards (Haury, 1976). The Coahuilla made baby "diapers" and women's skirts from mesquite bark which was rubbed and pounded until it softened (Hooper, 1920; also see Palmer, 1878).

Mesquite wood served most technological functions for which wood was required. It was extensively used for making weapons: war clubs, atlatls, bows, and fending sticks (Grossman, 1873; Russell, 1908; Hooper, 1920; Gifford, 1936; Cosgrove, 1947). However, more flexible woods, such as cat-claw (*Acacia greggii,* Leguminosae) or desert hackberry (*Celtis pallida,* Ulmaceae) were often preferred for bows (Felger and Moser, in prep.).

Prior to the introduction of steel, the Pima made awls from mesquite wood (Russell, 1908) and the Seri used large mesquite spines for sewing (Felger and Moser, in prep.). Planting sticks, digging sticks, weed cutters, and pestles were often made of mesquite (Russell, 1908; Forde, 1931; Cosgrove, 1947; Felger and Moser, in prep.). In the nineteenth and early twentieth centuries, the Pimans made shovels from solid pieces of mesquite wood (Russell, 1908). One of these shovels measures 133 cm in length, the blade being 43 cm long and 28 cm wide (Arizona State Museum, E-116). Trays made of mesquite wood were also prized household possessions of the Pima, and were used for such purposes as mixing bread (Russell, 1908).

A wooden bowl, usually made from mesquite, known as a *batea,* as well as spoons, have long been common kitchen items in northern Mexico (Thomas Hinton, personal communication). Use of mesquite for furniture, particularly for stools, table tops, and chairs, extends from pre-Columbian times until today in rural regions.

Uses for which the Seri traditionally have used mesquite wood include the carrying yoke, pestle for pounding mesquite pods and other hard fruit and seeds, clubs for killing fish and sea turtles, a part of the columnar cactus fruit-gathering device (Felger and Moser, 1974a), specific cooking forks, violin bows, and occasionally for the violin box (Felger and Moser, 1971; in prep.). The Maricopa made three kinds of paddles for pottery manufacture, all of mesquite wood. To seal a pottery vessel, they filled a still-hot and freshly fired pot with a fine gruel of ground mesquite pods (Spier, 1933).

Unless it is treated, mesquite wood is generally not cut during the summer months. However, it is easiest to cut during summer, from May until early September, because at that time the wood has the highest moisture content. Summer-cut wood invariably becomes riddled by a certain large wood-boring beetle larva, and the wood will be of little value unless treated. Summer-cut wood is sometimes stacked around a fire and heat sterilized. The logs are often slightly charred but the borers are killed. Another and often preferred method of curing summer-cut wood is to soak freshly cut wood in a pond for several weeks. The sap leaches into the water, producing an extremely unpleasant and strong odor and the water appears to be fermented.

The logs are then dried for about a month. Water-cured, summer-cut logs will be substantially lighter in weight than winter-cut wood, presumably because of the leaching of the sap. This is an important consideration if the wood is to be transported by pack animals (Alexander Russell, personal communication).

Since Spanish colonial times mesquite has been extensively utilized for fence posts and corrals. The traditional stacked-pole corral requires considerable skill and effort to construct. A large quantity of relatively straight logs about 1.5 to 2 m long must be obtained. Construction of a single 200-foot-square stacked-pole corral on the Papago reservation was estimated to have required seventy cords of mesquite logs (Clotts, 1915)! Although this would be a large corral, it is by no means unusual to still find corrals of this size in northern Sonora. Mesquite posts cut at the correct season or properly cured will last at least ten years after being set in the ground (Alexander Russell, personal communication).

Mesquite house posts, beams, roof supports, and smaller twigs have been extensively utilized in traditional architecture since ancient times (Haury, 1976). The corner posts and roof beams of recent Seri huts as well as the traditional ramada are commonly made of mesquite (Felger and Moser, 1971). Mesquite trunks are often crooked or forked and for this reason they were primary factors determining the low ceilings of many traditional native homes (Holden et al., 1936). Since Spanish colonial times, high ceilings have become more common, and use of mesquite for the main architectural members has waned. Yet Spanish colonial and territorial architecture makes extensive use of mesquite lintels over doors and windows, and occasionally for door posts and *vigas*. However, until recent decades mesquite door posts were commonly utilized in simple rural homes. Two posts were selected about 2.2 m long and to be as straight as possible. These were set at either side of the door, grooves cut in the posts, and a door was then fashioned from planks and hung on the post with hinges. Since the posts were seldom straight, the door had to be cut to fit the irregularities of the posts (Alexander Russell, personal communication). Forked mesquite posts were widely used to support large water-storage pottery vessels or *ollas* near the house (Spier, 1933).

Because it is water resistant and readily available, mesquite has been the preferred wood for the ribs and most of the framework of fishing boats and other small craft in northwestern Mexico since Spanish colonial times (León-Portilla, 1973; Felger and Moser, in prep.; Alexander Russell, personal communication). Curved limbs are individually selected to fit the design. However, mesquite wood is seldom suitable for the planking or keels because it does not yield long, straight-grained boards.

FUEL

Use of mesquite as a preferred as well as necessary energy source for cooking and heating has been widespread and is still popular among people of southwestern North America (Meigs, 1939; Felger and Moser, 1971). It imparts a good flavor to food and burns evenly and hot. Excavations indicate that mesquite-fuel hearths have been a commonplace feature in the Southwest since ancient times (Haury, 1950) even though it is seldom mentioned in the literature. In many parts of northwestern Mexico mesquite wood continues to be the most important energy source for cooking meals. For domestic cooking purposes there is a strong preference for wood less than about 10 cm in diameter, and in northern Mexico housewives will often not purchase wood of larger size. The occasional larger pieces are used primarily for heating. Mesquite wood collected for fuel can be either dry dead wood, or it can be cut green during the cooler times of the year, and stacked and dried for a number of months until sufficiently dry. While cured, greencut wood is often preferred by the housewife, price and availability dictate the kind and condition of wood available. Because it is salable immediately, the dry dead wood is usually removed by commercial woodcutters before live wood is cut. For this reason, green wood is often the only wood available in populated areas.

It may be presumed that aboriginal hearths were generally constructed to utilize relatively modest quantities of firewood. Among the Seri and other Southwest Indians the kinds of foods prepared in traditional times often required only modest cooking time. For example, seeds or grains and other plant-derived foods were generally ground into flour. By reducing food materials to small particles with a high surface-volume ratio, cooking time is markedly shortened, and such foods were commonly consumed as gruels (Felger and Moser, 1975). For these hearths dead dry wood suffices and small pieces of wood are utilized. Among the Seri and others, the gathering of firewood was woman's work (Felger and Moser, 1976).

Mesquite wood is suitable for most purposes requiring high, steady heat, such as firing pottery (Bell and Castetter, 1937; Balls, 1962; Fontana et al., 1962). More recently, it has even been used for such purposes as tempering steel drills for drilling rock (Alexander Russell, personal communication).

In Mexico, woodcutters, or *leñeros,* have long been regarded as being in a low-status occupation. In recent decades their horse-drawn wagons have largely been replaced by stake-sided trucks, often about 2 to 2 1/2 ton capacity, and their tracks extend into remote reaches of the desert. The woodcutters have often been responsible for establishing access into remote places in the Sonoran Desert. After completion of the paved highway (Mexico, Route 2) from Sonoyta west to San Luis, Rio Colorado, Sonora, in 1956, which replaced the infamous Camino del Diablo (Ives, 1964), woodcutters gained access to the Pinacate region (Hayden, 1967). "The Mexican woodcutters from San Luis who immediately began to search the arroyos in the lava flows for ironwood and mesquite laid down an ever-widening network of truck tracks, many of which followed Indian trails . . . All credit must be given to the woodcutters who have, with their dilapidated trucks, with a barrel of water, a sack of frijoles, and an abiding faith in God, laid down tracks where no sensible person would have driven an army tank" (Hayden, 1967:335).

Similarly, various Indians, such as the Apache (Hrdlička, 1908) and the Papago have often supplemented their incomes by wood cutting.

With shortages and associated price increases, wood cutting has become more profitable, particularly north of the Mexican border, with a cord of home-delivered mesquite wood fetching $50 to $60 or more in southern Arizona cities. Four cords of wood commonly constitutes one load of a modern two ton truck (Richard Crossin, personal communication). The present harvest of mesquite wood for luxury home heating and restaurants, as well as for everyday cooking on both sides of the border, is not being practiced on a sustained-yield basis, and the resource is diminishing. Since larger diameter wood is selected for heating rather than for cooking, particularly in the case of luxury usage, wood cut in the United States tends to be from older trees and limbs than that cut in Mexico, where the primary use is still for domestic cooking. Mesquite stumps readily produce new growth after cutting, and the regrowth wood is excellent for cooking fuel. Thus, particularly in the case of domestic cooking fuel, the resource would seem well suited for sustained-yield management practices.

RELIGION AND RITUAL

While mesquite was a staff of life in arid regions of southwestern North America, it was generally not holy or venerated. It was essential but not noble. Unlike cultivated crops, human intervention seldom seemed necessary to insure production of the crop, which was nearly always available in excessive quantity. Nevertheless, other dependable or unfailing perennial wild crops, such as the saguaro (*Carnegiea gigantea*) among the Papago (Castetter and Underhill, 1935), do figure into native religious practices. Saguaro fruit was the source of an intoxicating wine and food employed in a mystical sense to bring forth rain for the planting of crops, heralding the Sonoran Desert New Year.

Among most tribes, such as the Seri, information regarding ritualistic use of mesquite is negative. With the exception of the Cahuilla, the few known cases of ritualistic practices involving mesquite generally do not center on mesquite itself as the object of the ritual. Some of the more prominent cases are mentioned below. Among the Cahuilla, "religious sanction was required prior to gathering the pods, which was accomplished by means of a rite known as 'feeding the [ceremonial] house'" (Bean and Saubel, 1963:66), which served to bring together the members of a given lineage. The ceremonial or political leader picked fruit to be prepared and eaten in the ceremonial house, after which others were free to gather the crop. A special class of men responsible for certain religious practices were entrusted with powers to regulate rain and other natural phenomena and it was their duty to bring early spring rains which the Cahuilla regarded as necessary to insure an ample mesquite crop (summer rain is rare in the California deserts). Furthermore, at harvest time these men were responsible for holding off summer rains, so that dampness would not spoil the fruit (Bean and Saubel, 1963, 1972).

Various people, such as the Cahuilla, Kamia, Maricopa, Mohave, and Quechan, had clans or lineages named for the mesquite, some of which had totemic significance (Bean and Saubel, 1972; Bell and Castetter, 1937; Bourke, 1889; Gifford, 1918). Among the Maricopa a feathered pike of mesquite wood was carried into battle, and, since it was to be in the front of the battle, the warrior carrying it became the battle leader (Spier, 1933). A Pima warrior who killed an Apache underwent a cleansing ceremony involving plastering the hair with mesquite gum and black clay (Grossman, 1873). Similar practices were observed by the Maricopa, and during their sixteen-day purification rite the only food eaten was small quantities of mesquite *atole* or gruel (Spier, 1933).

Plasters of mesquite gum and clay were involved in a Quechan Indian girl's puberty ceremony, the plaster being applied to the girl's hair and also to the hair of men who happened to be visiting during the time of her ceremony (Curtis, 1908; Bolton, 1930; Forde, 1931).

The Papago sometimes buried their dead in a sitting position, and over the burial a roof was erected of palo verde (*Cercidium* spp.) or mesquite taken from the deceased man's house (Lumholtz, 1912). A green mesquite pole was placed on each side of the Cocopa funeral pyre (Gifford, 1933), and as early as the first and second century B.C. the Hohokam people of Snaketown in Southern Arizona cremated their dead on mesquite wood pyres (Haury, 1976).

SUMMARY

Mesquite was the most widespread and important resource of the diverse native peoples in southwestern North America. It was utilized for food, fuel, shelter, weapons, tools, fiber, dye, cosmetics, medicine, and a multitude of other practical as well as aesthetic purposes: every part of the plant was used. Mesquite and several other major perennial wild crops were predictably available for harvest at the height of the pre-summer dry season and at the onset of the short summer monsoon season, making this a time of plenty. European introduction of winter-spring agricultural crops, namely winter wheat, provided a substitute for the mesquite harvest.

The mesocarp of the pods provided a major carbohydrate or calorie-rich component in native diets. It was primarily prepared as flour, and commonly made into a gruel, cakes, and beverages. The seed was not extensively utilized in historical times even though it is high in protein content. However, a specialized tool, the gyratory crusher, appears to have been developed in ancient times for processing substantial quantitites of mesquite seeds. The whole pods, flour, and prepared cakes were commonly stored in large quantities for future use. The herbage and sap or gum feature prominantly in the regional pharmacopoeia, and the most common medicinal usage was for treating eye ailments.

Chapter 9

MESQUITE AND MODERN MAN
IN
SOUTHWESTERN NORTH AMERICA

C. E. Fisher

Native inhabitants, and later, early settlers in the American southwest considered mesquite invaluable as a source of fuel, fence posts, building material, utensils, weapons, and even food for both themselves and their animals. The shrubs or small to moderate sized trees also provided cover and shade for man, domestic animals, and wildlife in periods of high summer temperature, and refuge from severe winter storms. The fruits or beans, were used in the preparation of foods, beverages, and medicines, the gum exudate for making glue and candy, and individual trees were used as boundary markers (Bray, 1906; Forbes, 1895; Griffiths, 1904; Havard, 1895; Palmer, 1878).

At the turn of the century, after a period of thirty to fifty years of intensive land use and a striking buildup in the population of domestic grazing animals, *Prosopis* came to be considered a pest rather than an asset in the southwestern United States (Figure 9-2). Similar recent changes in population densities have been reported for *vinal* (*P. ruscifolia*) and other *Prosopis* species in Argentina (Morello et al., 1971) and in the Chaco region of Paraguay (Fisher, 1964), as discussed in Chapter 10. Following the introduction of livestock and other disruptive activities, natural populations of mesquite and several other species gradually became transformed from open groves of trees (Long, 1820; Marcy, 1849) to moderate or even dense thickets (Figure 9-1). Just below the surface of the soil in the case of young plants, in all species of *Prosopis* observed, and several centimeters below the soil in older plants, is a zone of preformed buds that remain dormant as long as the primary bud or tree continues normal growth and development. If the initial seedling growth or tree trunk is destroyed in any way, such as by freezing weather, drought, fire, cutting, or trampling, these underground buds initiate new growth around the base of the plants (Fisher et al., 1946). The resultant multistemmed regrowth often occupies a much larger basal area and may be more objectionable than the original plant. This change in growth form and the increased populations of *Prosopis* plants are largely responsible for great increase in the area now occupied by thorny, impenetrable brush thickets (Figure 9-3). This change is often accompanied by a decrease in the production and accessibility of more desirable forage plants.

THE SPREAD OF MESQUITE IN HISTORICAL TIMES

Although the surface coverage by *Prosopis* and other similarly growing trees has increased in the last seventy years, the limits of its geographical distribution have remained essentially the same for the last 150 years (Johnston, 1963; Malin, 1953). Some range expansion to new areas has occurred (Fisher, personal observation), however, along cattle trails and areas where the natural competition offered by associated species has been destroyed, but

FIGURE 9-1. *Early stages in the infestation of rangeland by* Prosopis glandulosa *following intensive grazing and disruptive utilization by modern man.*

FIGURE 9-2. *Invasion of the Santa Rita Experimental Range near Tucson, Arizona, by* Prosopis veluntina. *a. A view of the area in 1903. b. The same area in 1941 showing the increase in mesquite population densities.*

FIGURE 9-3. *Heavy infestation of* Prosopis glandulosa *in western Texas. The introduction of large numbers of grazing animals and exploitation of the former grassland by modern man has resulted in the increase in the frequency of multistemmed individuals of* Prosopis *throughout extensive areas of formerly highly prized rangeland.*

the major change has been due to increased population densities in areas in which *Prosopis* has always been found (Bray, 1904; Cook, 1908; Smith, 1899). The present geographical distribution of *Prosopis* in the southwestern United States (Figure 9-4) is estimated at about 22.7 million hectares (70 percent of the surface area of the state) in Texas, 1.62 million (10 percent) in Oklahoma, and 9 million (20 percent) in New Mexico and Arizona. It also occurs to a small extent in southern Utah and California (5 percent), and has become a major pest on extensive areas in central and northern provinces of Argentina (Fisher, 1962; Feldman, 1972), the Chaco region of Paraguay (Fisher, 1964), and the northern and western parts of Mexico (Rojas, personal communication).

Three species of *Prosopis* (*P. glandulosa, P. velutina,* and *P. pubescens,* see Appendix) have been recognized in the United States, but the only species of major concern are the velvet mesquite (*P. velutina*) in Arizona and the honey mesquite (*P. glandulosa*) in California, Arizona, New Mexico, Utah, and Texas (Figure 9-5a). In northern Argentina and the Chaco region of Paraguay, vinal (*P. ruscifolia*) occupies immense areas in almost pure stands (Figure 10-2) and is considered a national pest (Feldman, 1972). Other species that have spead rapidly under grazing and modification of the habitat by man include (*P. nigra*), itin (*P. kuntzii*), nandubay (*P. affinis*), *algarrobo parguayo*

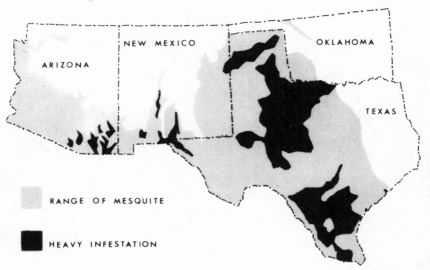

FIGURE 9-4. *The distribution of* Prosopis *in southwestern United States and areas where it has become a problem on grazing lands.*

(*P. hassleri*), *calden* (*P. caldenia,* Figure 9-6b) and others including various hybrids of the different species. Nearly all species can vary from large, single-stemmed trees to many-stemmed shrubs depending on climate, soil, density, and injury to the above-ground growth (Fisher et al., 1946). Once established, individual trees (at least of *P. velutina*) can develop a deep tap root that can reach to depths below 50 m (Phillips, 1963) and lateral roots extending to 13-15 m (Fisher et al., 1959). This dual root system is efficient under drought conditions and permits *Prosopis* to compete favorably with other species (see Chapter 2).

FACTORS INVOLVED IN THE SPREAD OF MESQUITE

Under normal conditions, there seems to have existed a quasi-stable balance between desert grassland and desert scrub that was upset by any one, or a combination of several factors (Parker and Martin, 1952). It is essentially the alteration of the factors that maintained grassland which resulted in a distinct change in growth form and increased densities of mesquite populations. These factors include the control by natural recurrent prairie fires, long-term drought, grazing by domestic livestock, and utilization by man. All have undoubtedly contributed to the destruction of the pre-1850 relationships of *Prosopis* and the native herbaceous vegetation. The elimination of natural periodic fires was formerly thought to have been the most important factor that encouraged the spread of *Prosopis* (Humphrey, 1949). However, within recent years, studies conducted in southern Arizona (Parker and Mar-

FIGURE 9-5. *Above: Dense coverage of* Prosopis glandulosa *and other weedy taxa such as* Opuntia *on formerly more open grassland in Texas. Below: The* calden, Prosopis caldenia, *has similarly increased in density in the Pampa, Argentina, the most important cattle-raising area of the country.*

FIGURE 9-6. *Control of previously dense stands of* Prosopis glandulosa *in Texas improves the production of range forage and restores highly productive range sites. The treatments used in this area of the western part of the state were root plowing, brush raking, and reseeding. Large trees such as* Quercus virginiana *seen in the background are considered desirable on this range site.*

tin, 1952) and in northwestern Texas (Fisher, 1947; Robison, 1967; Wright, 1973) have shown that under present grassland conditions, fire seldom kills more than a few small plants of *Prosopis.* Undoubtedly, occasional heavy accumulations of litter prior to the introduction of livestock may have furnished a large enough supply of flammable material to generate heat sufficient for widespread destruction of many small *Prosopis* plants. Nevertheless, the absence of grass fires in modern times does not alone seem to be sufficient explanation for the dramatic change in the growth form and densities of *Prosopis* populations since 1900 (Hastings and Turner, 1965).

The rapid influx of large numbers of grazing animals into Texas (from fewer than one-half million head in 1830 to over 9 million in 1900) (Texas Almanac 1857, 1973) was very likely the most important single factor influencing the increased density of mesquite. In New Mexico, the number of cattle increased from about 158,000 in 1870 to over a million in 1886 (Ares, 1974). During the same period, the land surface covered by mesquite on the Jornada Experimental Range increased from 4.8 percent to 50.3 percent while grassland declined from 90 percent to 25 percent (Buffington and

Herbel, 1965). The advent of the windmill and stock tanks that provide water for livestock further aggravated the problem by extending the area that could be intensively grazed over long periods of time (Fisher, 1950). In Argentina, Feldman (1972) postulated that a combination of heavy exploitation of large desirable trees of *Prosopis* and associated species, increased grazing pressure brought about by the development of agriculture and, perhaps, a change from sheep to cattle were some of the major factors responsible for the spread of many species of algarrobos. In the Chaco region of Paraguay, introduction of large numbers of domestic grazing animals made possible by the development of watering facilities, roads, and control of predators was closely followed by invasion by vinal (*P. ruscifolia*) and other species of *Prosopis* (Fisher, 1964; see Chapter 10).

The role played by grazing animals in accelerating the increase of *Prosopis* is quite clear. Domestic livestock, wildlife, and rodents relish the highly nutritious pods (see Chapter 7). The seeds are long-lived (Martin and Tshirley, 1962) and germinate readily after passing through the digestive tract of most grazing animals (except small rodents) and are disseminated in their droppings (Fisher et al., 1959; see Chapter 2).

Great opportunity for dispersal of the seed, heavier stressful utilization of normally competitive range forage plants, and severe disturbance of the soil by trampling in areas where large numbers of grazing animals congregate offer almost ideal conditions for rapid invasion by *Prosopis*. Some of the heaviest infestations in the early days of grazing and during recent times have developed around areas such as roundup grounds, trails, and along drainage ways where forage cover has been depleted and erosion of the soil has taken place.

Recurrent droughts probably added to the impact of heavy range use in upsetting the balance between *Prosopis* and the native herbaceous plants. It has been shown (Scifres et al., 1971) that large numbers of *Prosopis* germinated and became established following the severe drought in the Rolling Plains area of Texas. Repeated observations also have shown that natural stands of *Prosopis* are composed of distinct size and, presumably, age classes that may have become established during periods of severe drought when the cover of herbaceous plants deteriorated.

Another factor that might have aided the increase of *Prosopis* population sizes may have been an alteration of its natural relationships with parasitic insects or predaceous mammals (primarily rodents) and birds (see Chapter 7). Other than the extermination of large numbers of the prairie dog (*Cynomys ludovicianus*), known to locally control seedlings and young plants of *Prosopis*, there is no evidence that any major change in wildlife population sizes or activities have occurred that could have influenced the spread of mesquite.

Many ranchers have implied in the past that *Prosopis* actively destroys grassland rather than invading areas only after the natural herbaceous cover has been altered by grazing animals. The effect that *Prosopis* may have on the growth of perennial grasses and other native grasses and other native forage species is under study, but is not easily determined. The degree of impact of mesquite on herbaceous plants seems to result from a combination of factors such as the population density and growth form of *Prosopis*, the supply of available soil moisture, soil type, and the original kind of vegetation cover. Under good soil moisture conditions and a natural cover of vegetation in the northwestern areas of Texas, low to moderate densities of mesquite have few harmful effects on the herbaceous yield (Dahl et al., 1973). Under somewhat less favorable moisture conditions (with an average rainfall of 54.2 cm/year) at Spur, Texas, the removal of over 1,800 *Prosopis* trees and shrubs per hectare did allow a 32 percent increase in herbaceous plant growth over an eight-year period (Robison et al., 1970). On the other hand, where rainfall tends to be sparse and erratic, as in southern Arizona, the removal of 249 *Prosopis* trees per hactare prompted a two-fold increase in herbage by essentially allowing an increase in the soil moisture content to a depth of 15.2 meters (Parker and Martin, 1952). In a later study it is stated that, under limited rainfall in Arizona, velvet mesquite robs the soil between the crowns of moisture and nutrients, but compensates in part by providing a more favorable environment beneath the tree crowns (Martin and Cable, 1974).

CONTROL OF MESQUITE IN THE U.S.A.

These findings strongly indicate that the amount of soil moisture, soil type, and *Prosopis* population densities all may affect the growth and production of perennial range forage plants. However, it should be remembered that these factors are correlated with one another and that both the state of the present herbaceous vegetation and mesquite populations are "artifacts" produced by man's activities. Today, the control of *Prosopis* on grazing lands has become one of the major agricultural problems of the American southwest (Fisher et al., 1959; Norris et al., 1963; Herbel, 1969; Parker and Martin, 1952). Research to develop both effective and economical methods to control the increased population sizes of *Prosopis* on grazing lands has been underway for many years by state and federal organizations working in close cooperation with private industry and ranchers. Many methods of control have been suggested and tested. To be useful on a large scale, a control method must be relatively low in cost, effectively remove or destroy established stands, and prevent rapid reinvasion of the treated areas by either seedlings or sprout growth from partially killed plants. Fire or mechanical and chemical treatments that kill only the above ground growth lead to subsequent heavy, multistemmed regrowth from the trunk base. In order to prevent resprouting, it is necessary either to remove or destroy the lowest dorman buds on the underground portion of the stem.

Fire is not currently used to control *Prosopis* for two reasons. First, there is now insufficient flammable material on the ground to produce a fire, intensive enough to effectively kill established *Prosopis* trees. Second, this method often leads to a concomitant loss of many desirable forage plants (Wright, 1973).

Application of 2,4,5-T (2,4,5-trichlorophenoxyacetic acid) in an oil solution to the trunk base or cut surfaces curbs the growth of individual *Prosopis* plants. Kerosene and diesel oil alone are effective if applied in amounts sufficient to penetrate down to the lowest dormant buds. Fenuron, a substituted urea compound, has been effective when applied either in water or in granual form around the base of plants (Fisher et al., 1959). Aerial applications of 2,4,5-T over extensive regions of almost impenetrable thickets has brought *Prosopis* under control (Figure 9-6). More recently, combinations of 2,4,5-T and picloran (4-amino-3,5,6-trichloropicolinic acid) have been found to be more effective than either alone and also effectively to curb other invading species (Robison, 1968; Robison et al., 1970). Retreatment to control re-invasion is usually required at intervals of five to seven years (Fisher et al., 1972).

Mechanical methods used extensively to control *Prosopis* include huge crawler tractors equipped with a blade to uproot individual plants, root plows that sever the roots, and anchor chains that knock down and tear up shrubs (Fisher et al., 1973b). Mowing, roller chopping, and shredding, all of which destroy only the above ground growth, are not effective methods when used alone.

Over the past decade in Texas, over 1.2 million hectares of rangeland infested with *Prosopis* have been treated annually with herbicides and mechanical equipment in order to keep encroachment by mesquite under control.

Despite the fact that *Prosopis* has become a problem due to modern man's ranch management practices, it continues to play a positive role on grazing lands. When properly controlled, mesquite provides food and shelter for livestock, a source of nectar and pollen for honey bees (Chapter 5), and a favorable environment in which wild animals can persist (Chapter 7). In addition, one of the major beneficial effects of *Prosopis* in areas that are grazed is that the shrub canopies provide shelter and enable palatable, desirable plants growing under them to avoid being eaten or trampled (Figure 9-7). They reproduce, and thus are able to persist under conditions of heavy grazing, high temperatures, and prolonged drought (Chapter 7). Survival of individuals of key species of perennial forage plants, *Buchloe dadyloides*, buffalo grass; *Bouteloua curtipendula*, sideoats grass; *Stipa leucotricha*, Texas wintergrass; *Panicum obtusum*, vine mesquite; *Trichachne californica*, Arizona cottontop; *Setaria macrostachya*, plains bristlegrass; *Pappophorum bicolor*, pink pappus, and others under a protective cover of *Prosopis* shrubs provides a seed stock of native adapted species that are continually needed to restore and improve grazing lands. Moreover, on moderately grazed land, the organic matter, total nitrogen, and porosity of the surface soils are higher under the canopy of *Prosopis* than in open areas and thus amenable for herbaceous growth (Tiedemann et al., 1971; Brock, 1972).

We can thus see that many of the features of the life history and physiology of *Prosopis*, under natural conditions that allow it to grow marginally in desert scrub environments (Chapter 1), accelerated its expansion once man altered the ecosystem. These features include the abundance of long-lived seeds that are disseminated by vertebrates, high germination rates over a wide range of temperature and moisture conditions once the seed has been freed from the endocarp, rapid development of a deep tap root system, tolerance to drought and grazing, and the ability to regenerate following injury (Glendening and Paulsen, 1955). With the introduction of cattle and the heavy utilization of native herbaceous plants, man simultaneously removed potential plant competitors, decreased the frequency of fires, provided an abundant agent for free seed dispersal, and frequently provided soil conditions favorable for *Prosopis* growth. In addition, the practically continuous injury to shrubs while attempting to utilize and control *Prosopis* invoked the regeneration response with the result that entire regions became covered with shrubby thickets rather than single to few-stemmed, tall trees. It seems inevitable in areas of this adaptation where rainfall and soil water relationships are relatively favorable such as in west Texas, *that Prosopis* will continue to be a problem on rangeland utilized by cattle and other grazing animals. However, judicious grazing practices in combination with shrub control should allow the modern rancher to combat the detrimental effects of *Prosopis* and still benefit from the presence of scattered clumps of trees on his rangeland.

SUMMARY

Within the last seventy years, the human view of mesquite has changed. In pre-European times, mesquite was a necessary and respected part of the lives of native Americans in the southwestern deserts. It is now considered the major agricultural pest of the southern rangelands. Several factors have contributed to the increased densities of *Prosopis* populations in these areas and the concommitant loss of agriculturally valuable grazing land. Among the most important of these factors are the cessation of range fires, the reduction of the natural grass cover, the increased dessimination of seeds by large herbivores, and the lessening of the effectiveness of invertebrate seed predators. Most of these factors can be directly or indirectly related to the activities of European man. The settlement of the Southwest by Europeans stopped the native Amerind practices of burning the grasslands. Historically, ranchers tended to overgraze the land which resulted in the removal of the grass cover. The large number of introduced cattle provided a highly effective seed dispersal agent that simultaneously carried the seed from the parent tree, removed the seed from the endocarp, and destroyed internal seed predators. With the advent of all of these factors beneficial to the dispersion and successful establishment of mesquite, population densities rose rapidly.

FIGURE 9-7. *The luxuriant growth of highly palatable grasses following the control of* Prosopis *in Texas shows the beneficial effect of limited mesquite growth. Occasional shrubs of* Prosopis *protect the grasses and forbs from heavy trampling and provide shade from high temperatures.*

Several decades of research on mesquite control have shown that judicious range practices can restore the natural grasslands and even make use of the favorable properties of scattered individuals of mesquite on grazing lands. The most effective control results from a combination of mechanical removal and herbicides followed by reduced grazing pressure.

As shown in the next chapter, the same reversal of the roles of mesquite in native versus introduced cultures occurred in Argentina where *Prosopis* played an important beneficial role in pre-Hispanic cultures but has now become regarded as a serious agricultural nuisance by European settlers.

FIGURE 10-1. *A branch of* Prosopis ruscifolia, *the* vinal, *a species which has become a rangeland pest in northeastern Argentina. The spines, borne singly at each node, can reach 35 cm in length.*

Chapter 10

ALGARROBOS IN SOUTH AMERICAN
CULTURES
PAST AND PRESENT

H. L. D'Antoni

O. T. Solbrig

Algarrobos have been economically important in South America since pre-Colombian times and today are still an important forage and timber resource. In the past, *Prosopis* fruits were a major food for humans as well as domesticated animals, but their present consumption by man is marginal and mostly traditional. However, the wood of various species of *Prosopis* is commonly used as fuel and in the construction of primitive dwellings, modern flooring, and fence posts. In recent times, one species, *P. ruscifolia*, known as the *vinal*, has become a serious agricultural problem in the Argentine states of Formosa, Chaco, and Santiago del Estero, because it aggressively invades pasture lands.

Parallel to the treatment of mesquite in the North American southwest (Chapters 8 and 9), we examine here in some detail the uses of *Prosopis* by ancient and modern South American cultures and indicate how and why the vinal has spread in both area and population sizes in Argentina in recent times. We will primarily be concerned with the Argentine *Prosopis* species and will indicate specifically uses for the species found near our Andalgalá study site.

PROSOPIS IN PRE-COLOMBIAN AND EARLY SPANISH SETTLEMENTS*

When the Spanish conquistadores invaded northern Chile and northwestern Argentina, they encountered trees with curiously contorted trunks and sweet, edible fruits. They found that these trees grew sporadically from the dry deserts of the Peruvian coasts to the more mesic areas of the region of Cuyo and the province of Córdoba in Argentina. Because the trees resembled the "algarrobo" (*Ceratonia siliqua*) of the Iberian Peninsula in their aspect and potential uses, the conquistadores gave this Spanish name to various species of *Prosopis*. The fruit pods of *Prosopis* were often depended upon as a food source, as when Almagro crossed the Atacama Desert in his exploration of Chile. His troops were sustained for a long period on the algarrobo fruits they were able to collect during the march. Oviedo y Valdés (1535) said in this respect ". . . *aquellos pueblos destas algarrobas que alli habia recogidas en cantidad, se hizo de ellas miel y pan para sostener a la gente . . .*" (". . . in these villages there were many of these algarrobo fruits from which honey and bread was made to sustain the people . . ."). It is known that Almagro's expedition suffered from starvation on their return trip. Oviedo y Valdés (1535) explained the differences between the initial and return marches by stating that the algarrobo fruit crop of the previous year was gone and the new crop had not yet matured when they later went north. For forty days, the troops lived on ten previously collected algarrobo fruits a day per person; an allotment each had to share with his horse.

The South American Indians, like the North American Amerinds, have a well-developed lexicon for referring to different parts of a *Prosopis* tree, but

*H. L. D'Antoni.

differ in having names for different species of algarrobos. For example, the Quechua (a linguistic group of Indians of the Peruvian highlands, Figure 10-2) name for *Prosopis* is *thaccu* or *taco* which means "the tree" or "the one," referring, no doubt, to the fact that even during the years of greatest drought, the trees yield fruits. In other areas of South America, parts of the year during which *Prosopis* fruits are ripe are named for the presence of algarrobo fruits. In the Vilela language (Figure 10-2) various species of *Prosopis* have distinct names: *malumpe, P. alba; nabise* and *mabise kirimit, P. nigra; nabis mop* (the grandfather of the black algarrobo, *P. nigra*), *P. affinis; lire bos* or *kire bun, P. kuntzei;* and *uinesh, P. ruscifolia* (Martinez Crovetto, 1965).

There are many references to *Prosopis* and the use of various species in the early Spanish chronicles. Father Bernabe Cobo (1653) wrote:

In Peru [there are] five or six species of trees known as guarango which produce edible fruits similar to the algarrobos [of Spain].

FIGURE 10-2. *General locations of the Indian tribes and archaeological sites of South American natives using* Prosopis. *Because of the scale of the base map, some of the tribal ranges appear larger than they actually are. More precise circumscriptions of the domains of South American Indian tribes can be found in Steward (1950). The stippled area is the phytogeographic region known as the Chaco.*

The Spanish call the one that produces the best fruits by the name of 'algarrobo de las Indias,' but it is different from the one in Spain. The fruits of the guarango are good eating and the Indians in some places [Tucumán, Argentina, and Paraguay] make flour and bread from them. In some areas the natives have little other sustenance than these fruits. The valleys where the algarrobos or guarangos are most abundant are those of Ica, Nazca, Guambacho, Casma, Chicama, Guadalupe and Catacos.

Oviedo y Valdés (1535) mentioned that the Atacama Indians ate corn (maize) for six months of the year and depended on *Prosopis* fruits to sustain them for the rest of the year. In the region of Cuyo (Argentina), the natives subsisted on algarrobo fruits and hunting (Lopez de Velazco, in Latcham, 1936). Likewise, the region circumscribed by the Mendoza, Desaguadero, and Tunuyán Rivers (Argentina) was rich in populations of *Prosopis* that provided fruits made into bread and chicha (an alcoholic beverage) by the Indians. Father Cabrera (1929) mentioned that the name *algarroberos* was given by the Spanish to these Indians because they periodically abandoned the fields they were cultivating for the Europeans in order to harvest *Prosopis* fruits.

As in the case of the North American mesquites, the uses of *Prosopis* by primitive and modern peoples can be divided into three principal categories: as food, as medicine, as fuel and timber. Although the literature on the South American uses of *Prosopis* is less complete than that for North America, I will briefly discuss some of the ways *Prosopis* is used within each of these categories. The areas inhabited by the native Indian tribes and anthropological sites mentioned in the discussion are shown in Figure 10-2.

Food

The fruits of *Prosopis* have played an important role in the diets of various native peoples of South America as far back as cultures of the dry areas of the continent can be traced. Towle (1961) has identified *Prosopis* beans in the excavations of Cahuachi and Huaca del Loro in the Nazca Valley of Perú dating about 1,200 years before present. Some drawings in the Mochica ceramics have been identified as depicting *Prosopis juliflora* but Towle (1961) thinks that this and other systematic identifications are incorrect. She differs, however, only in the specific identification and feels that the figures represent *P. chilensis* and not *P. juliflora*. These two species and *P. pallida* are very similar morphologically and difficult to tell apart even in the live condition. The fact that a species of *Prosopis* was represented has not been questioned.

In addition to Perú, *Prosopis* has left imprints on several other South American cultures. Many tribes of the Chaco (Argentina) had the custom of grinding together algarrobo fruits with those of *tusca (Acacia aroma,* Leguminosae) and *mistol* (*Ziziphus mistol,* Rhamnaceae). The resultant flour was mixed with water and the porridge-like mass put in an earthenware pot. The pot was placed in the center of a circle of seated individuals and each person helped himself to a handful of porridge from which he separated the non-edible parts (seed coats and fibers) and then ate the remainder. In contrast, the Ashlushlay, the Lengua, the Pilagá and the Mbaya (also perhaps other Chaco Indians, see Figure 10-2) made cakes of algarrobo flour that was mixed with water and baked (Métraux, 1946).

In reference to the Indians of northwestern Argentina, Lorano (in Heredia, 1968) stated that (in translation) "there are an infinite number of algarrobos some of which are called *zorruna* and which are not eaten but given to the animals, and others which are edible . . . of two species, one is quite juicy another white and drier. Both are ground and from the flour, bread is made which in Tucumán (Argentina) is called *patay* . . . they also make wine from algarrobos which is called *chica.* . . ."

To make *patay,* whole beans are collected and dried, then ground and sieved to remove the fibrous endocarp. The flour that results is mixed with water and formed into prismatic loaves which are dried and cooked in an oven. *Patay* probably formed one of the most important food stuffs made from *Prosopis* and is said (Burkart, 1952) to consist of 10–12 percent water, 4–6 percent fiber, 0.8–4.3 percent protein, and 55–65 percent carbohydrates (44 percent sugar and 11 percent starch plus cellulose). In addition, *patay* is known to have a high calcium content, a factor that makes it important in the nutrition of children. The presence of large amounts of calcium helps to explain also why there are few diseases associated with hypocalcemia in dry regions of Argentina even when there are periods of severe famine. *Churingo* is a special form of *patay* characteristic of the state of La Rioja in Argentina that is an uncooked mixture of *Prosopis* flour and water.

As recently as 1929 the inhabitants of the Huanacache lakes in Mendoza (Argentina) ate *patay* commonly. Similar to the Spanish designation of the Peruvian Indians, local inhabitants of this region became known as *algarroberos* (Métraux, 1929). Even today, *patay* is sold in the general stores in Andalgalá which is made by local people from the fruits of *P. chilensis* and *P. flexuosa.*

The inhabitants of the Chaco highly prized the fruits of algarrobos because the native peoples depended on them in former times for sustenance during part of the year. Métraux (1946) noted that between November and January, and occasionally even into February, the Indians in the region of the Pilcomayo River used algarrobos (mostly in the form of a beer) supplemented by the fruits of *chañar* (*Geoffrea decorticans,* Leguminosae) and mistol. Today, the Indians remaining in the Chaco search the forests during November and December for algarrobo fruits that are their source of sweets. The great importance *Prosopis* has had in the culture of these peoples can be deduced from the fact that one season of their five season calender

FIGURE 10-3. *Pilagá Indian women of the Chaco region sort algarrobo pods which they will grind into flour or use for making beverages. Photograph from Plate 49 in Métraux (1946).*

(November to January) is called the "season of algarrobo, chañar and mistol" (Métraux, 1946). The season of the algarrobo harvest was the time of joy and visiting, in contrast to the season of fishing when the majority of fights took place.

Many other groups of Indians also depended on *Prosopis* as a food source to a lesser degree. Lothrop (1946) mentioned that some tribes such as the Chaná, which had no agriculture, depended in part on wild fruits such as *Prosopis,* for food. The Comechingones (Aparicio, 1946), Humahuaca, located north of the Comechingones (Casanova, 1946), and Mocovi (Martinez Crovetto, 1968), although possessing agriculture, supplemented their diets with algarrobo fruits. Even in the Chaco-Santiago culture based on agriculture, livestock and the gathering of wild fruits, *Prosopis* also figured prominently (Figure 10-3).

In addition to *patay, Prosopis* fruits were often used to make beverages. A simple mixture of flour from dry ground pods and a large amount of water yields a drink known simply as *miel* (honey), while *añapa* is a beverage made from grinding fresh pods with water and straining out the pulp (Burkart, 1952). Finally, an alcoholic drink, *aloja,* is made by fermenting a mixture obtained by mashing one part dry fruits with four parts water. After twenty-four hours, the mixture begins to ferment and after forty-eight hours the pulpy part is removed. Today, *Prosopis* fruits are often used as substitutes for European products for which a taste was acquired after the conquest. One such use is the preparation of "coffee" from an infusion of toasted, ground beans. The true algarrobo, *Ceratonia siliquea* of Spain, was used in a similar way. Finally, algarrobo flour was used to form a covering layer over *bolancho*, a past made from *mistol* (*Ziziphus mistol*) flour that was eaten in the province of Santiago del Estero, Argentina (Burkart, 1952).

Medicinal Uses

Fernando Márquez Miranda (1946) mentioned various therapeutical uses of the algarrobo such as the use of fresh buds of *Prosopis nigra*, the black algarrobo, soaked in water with boric acid as a cure for conjunctivitis. Leaves of some of the South American taxa were mashed with animal fat to make a paste that was believed to help mend broken bones. Preparations of leaves were also considered good preventatives of didropesia, liver stones, or dyspepsia. A tea brewed from the leaves of *Prosopis strombulifera* has been used to treat venereal disease.

Timber and Fuel

Wood of *Prosopis* was used by ancient peoples of the arid regions in construction of primitive dwellings in South América as in North America. Algarrobo wood has been found in the tombs of Nazca and at other archeological sites in Perú (Yacovleff and Herrera, 1934, 1935) and was used in the construction of Huaca Prieta del Guanape (Figure 10-2) dating from the Cerro Prieto Period about 2500 B.C. (Bird, 1948; Strong and Evans, 1952). Huaca Prieta, located on the northern Peruvian Coast, is the earliest known site identified with the Upper Archaic Peruvian cultures. The site consists of an artificial mound with refuse piles and ruined houses (Lumbreras, 1974).

PROSOPIS AND MODERN MAN*

The modern uses of *Prosopis* in South America are primarily as forage and fuel or as lumber for the construction of primitive buildings, fence posts, pavements, or floorings. As an economic resource *Prosopis* is of importance primarily in Argentina and to a lesser extent in Paraguay, Chile, and Perú. This sequence of importance is, of course, partially correlated with the fact that Argentina has the largest number of species of *Prosopis* and a greater development of the genus than any other country (see Appendix). The discussion of the modern uses of *Prosopis* in South America will therefore refer primarily to Argentina, and, when possible, to the uses of the algarrobos of the dense bosque of *Prosopis flexuosa* that encircles the salar at the bottom of the Bolsón (Figure 1-6) near our study site at Andalgalá.

*O. T. Solbrig.

In Argentina, natural forests of algarrobos are exploited without regard to reforestation. Wholesale destruction of these woodlands has led to the decimation of *Prosopis* forests throughout northern and western Argentina. Destruction of stands of algarrobos occurred especially during the two world wars when shortages of imported coal led to the use of *Prosopis* wood (primarily *P. caldenia*) as fuel for steam locomotives and industrial furnaces. Coupled with the expansion of agriculture and cattle ranching, logging has reduced the natural coverage of algarrobo forests to about one-half or one-quarter of their former surface area. Furthermore, even where the forests have not been totally destroyed, the largest trees have usually been removed. This selective removal can be seen in the Bolsón de Pipanaco where there are few large trees of *Prosopis flexuosa* extant. During the last part of the last century and the first part of the twentieth century, the *Prosopis* forest in the Bolsón was used to furnish timber for the tunnels in the Capillitas copper mine and as fuel for the smelting ovens in the town of Pilciao, south of Andalgalá (Vervoorst, unpubl.). The modern, large-scale destruction of the *Prosopis* forests contrasts strongly with the sustained yield use of these species carried on for centuries by native Indians before the arrival of European man.

The wood of *Prosopis* is heavy and dense, with a high tannin content (Tortorelli, 1956) and consequently makes a good firewood that burns with a high caloric value. The primary modern use of *Prosopis* wood is, in fact, as fuel. In the past it was used for steam locomotives and is today still in great demand as fuel for cooking stoves and for heating. The species most commonly cut for fuel are *P. alba, P. affinis, P. caldenia, P. chilensis, P. kuntzei, P. nigra, P. ruscifolia,* and *P. torquata.*

In addition to its denseness, *Prosopis* wood has short, abundant wood elements that give the wood a high tensile strength important in some types of construction. The wood is actually not very different from oak wood, but the twisted short trunks of *Prosopis* trees preclude the use of algarrobo wood for any purpose requiring large pieces of lumber. The wood is used to make railroad ties, beams, and uprights in primitive dwellings and for slats in wagons and barrels. Fence posts of *ñandubay* (*P. affinis*) wood are highly prized as are those made from the wood of *P. torquata.* The very hard wood of *Prosopis kuntzei* is even used to make turnery articles.

A former heavy use of *Prosopis* wood in Argentina, but one which is now fading, is as material for the paving of roads. Cubes, 10 cm on each side, were used extensively instead of cobblestones to pave streets. The cubes were set on a base of sand and then covered with a light coat of tar. The surface produced is very smooth, absorbs street noises well and lasts ten to fifteen years. All of the major avenues of downtown Buenos Aires and many other large cities in Argentina were paved this way up to thirty years ago. Algarrobo cobblestones are known as *tarugos* in Spanish and were made primarily from the wood of *P. alba* and *P. nigra.* High labor costs and the increasingly heavier weights of vehicles have practically eliminated this type of pavement. However, *Prosopis* wood is still extensively used for making floor tiles. These tiles are arranged in patterns to produce parquet floors that are in great demand in Buenos Aires and other large cities. The principal species lumbered for this use is *P. caldenia*, the *caldén*, but trees of *P. flexuosa* near our study area in Andalgalá are used by a local parquet factory (Figure 10-4).

One species of *Prosopis* in particular, *P. ruscifolia*, the *vinal*, has become a serious agricultural pest during the last thirty to forty years, especially in the Argentine states of Formosa, Chaco, and Santiago del Estero (Figure 10-5). The agricultural problems associated with this species strikingly parallel those of *P. glandulosa* in Texas (Chapter 9) both in their causes and effects. The foliage of this species is unpalatable to cattle although the fruits, similar in appearance to those of *P. flexuosa* (Figure 1-11), make good forage for livestock. Adult plants possess sharp spines up to 35 cm long (Figure 10-1) and 1 to 2 cm wide. Formerly more scattered, individuals of *P. ruscifolia* now form dense thickets that are impenetrable to cattle and men. Cattle eat the fruits from the edges of the thickets and deposit the seeds in their feces, thereby furthering the spread of the species. Unlike *Prosopis glandulosa* and *P. velutina* in the United States southwest, *P. ruscifolia* has actually extended its range considerably during the last forty years as well as increasing the density of existing populations. The species has taken over millions of acres of formerly fertile pasture land in the state of Formosa and parts of the Chaco and Santiago del Estero.

According to Morello (1972), the reasons for the buildup of populations of the *vinal* are as follows. The *vinal* is a pioneer forest species that naturally invades grasslands in the subtropical savannas of the Chaco. Live-

FIGURE 10-4. *Logs of* Prosopis flexuosa *at Pipanaco, a town south of Andalgalá in the Bolsón de Pipanaco. Most of the wood from these logs will go into the manufacture of tiles for parquet floors but some will be used as fire wood or for the production of charcoal.*

FIGURE 10-5. *Dense thickets of* Prosopis ruscifolia *such as shown in this figure became more numerous in the Chaco of Argentina as overgrazing caused a depletion of the natural vegetation cover. Poor land use by ranchers combined with several years of drought rapidly and severely reduced the area formerly covered by palatable grasses and forbs.*

stock, especially cattle and horses were introduced into these areas by the Spaniards, but the high temperatures and the poor quality of the grass made the region marginal for livestock and herds of animals were not initially numerous. To improve the quality of the grass, the Spanish settlers continued the old Indian practice of burning the savannas. The yearly burning of the grasslands eliminated the seedlings of woody plants and maintained, or even extended, the savanna grassland during the early periods of colonization. Since the grassland in this area is a stage in the normal successional sequence from marsh to subtropical forest, the use of fire by the settlers had the effect of retaining a mid-successional community (Morello et al., 1971). However, this successful practice came to an end during the ten years from 1930 to 1940 because of two factors, one climatic, the other man-caused.

The decade of the thirties was a period of extreme drought in northern Argentina. The drought favored woody colonizing plant species such as *P. ruscifolia* that require well-aerated soils. At the same time that the drought was reducing the grass available for livestock, the Chaco War broke out between Paraguay and Bolivia. The war produced an economic boom in Argentina because the armies of both warring countries clandestinely bought cattle

and horses from settlers in the border regions of the Argentine state of Formosa. As a result, the settlers increased their livestock populations by in-
troducing animals from other parts of the country. The excessive animal
populations resulted in a destruction of the grass cover while, at the same
time, large movements of animals promoted the spread of the seeds of *P.
ruscifolia*. When fires were set to renew the grassland, there was often insufficient fuel to sustain fires hot enough to destroy the woody colonizing
plants. Consequently, plants of *P. ruscifolia* became firmly established in the
former grassland. After the war, many of the newly settled people remained
in Formosa and put up fences and turned their attention to agriculture. Herds
of animals were allowed to remain large and the grass was never allowed to
recover. In addition, the increased human occupation precluded the use of
uncontrolled burning as an agronomical tool. Under these conditions, the
vinal continues to spread.

At the present time, a massive program combining mechanical treatment
(bulldozing) with herbicides is underway to curb the spread of the *vinal*.
However, as pointed out by Morello (1972), as long as the neighboring lowlands continue to be drained and the savannas overgrazed, the loss of the
grassland and the extension of the thorn forest is inevitable.

SUMMARY

In South America as in North America, *Prosopis* has played an important
role in the lives of both the original native peoples and the modern inhabitants. The Indian peoples of the temperate and subtropical arid regions did
not cultivate algarrobos, but made abundant use of the pods as a food source
and of the flowers and leaves as pharmaceutical ingredients. The wood of
Prosopis has always been extensively used for fuel and simple construction
purposes. In general, *Prosopis* has not become a pest in the desert scrub areas
of western Argentina as it has in comparable areas in southwestern North
America. Part of the failure of *Prosopis* to establish itself in areas away from
washes in the Monte Desert could be due to the pattern of rainfall that precludes seedling establishment in any area except along river courses. The area
in which *Prosopis* has spread and become an agricultural pest in South America is in the Chaco of northeastern Argentina and adjacent Paraguay. Here,
overgrazing combined with the aggressive nature of *vinal, Prosopis ruscifolia,* and an unfortunate ten-year period during which the *vinal* was allowed to establish itself, has led to the production of forbidding thickets
covering enormous acreages of land. In general, *Prosopis* is considered a valuable tree in South America and is still highly prized for its wood, considered to
produce excellent floors and to be a superior fuel. In areas of western Argentina, some experimentation is being carried out to determine the feasibility
of large-scale use of *Prosopis* fruits as forage.

APPENDIX: The Genus *Prosopis* and Annotated Key to the Species of the World

A. Burkart and B. B. Simpson

Like nearly all members of the Leguminosae, species of *Prosopis* produce their seeds in legumes or pods (Figure 6-2), though unlike many other legumes, the mesquite pods do not split open at maturity. Differences in the shape of the pod and the arrangements of seeds inside the fruit provide characters that are common to groups of species. Relative proportions of the pod, the number of coils, and the degree of flattening are usually characters of individual species. Within the Leguminosae, *Prosopis* belongs to the Mimosoideae and shares with the other genera of this subfamily the characters of small flowers clustered into inflorescences (Figure 5-4), pinnately compound leaves with stipules (Figures 1-11, 3-3), and regular flowers with a five-parted corolla (Figure 5-5). Throughout the genus, there is great variation in the origin and form of the spines. Some species appear to be consistently spineless while others have occasional spineless individuals. In most cases however, spines are a prominent feature of *Prosopis* trees and shrubs. The spines can be formed by modified axillary branches, stipules, branch tips, or localized cells along a branch. Spines formed by axillary branches can occur in pairs (A-1), or singly (A-14) above a leaf node. Those produced by modified stipules occur only in pairs, but always below a leaf node (A-17). Spiny modified branch tips can appear segmented (A-15) because of the presence of leaf scars, or look lined (striate) (A-13). A few species have prickles (A-23) formed by localized cells scattered along the branches. The compound leaves (Figure 3-3) bear leaflets that vary in shape (A-3 versus A-4), in the width of the spaces between them (Figure 1-11), and relative size (A-6 versus A-7). Inflorescences can be ball-shaped (A-21), in the form of an open, relatively stubby, bottle brush (A-16), or densely flowered, long, slender catkins or spikes (A-5).

The species of *Prosopis* can be distinguished from other mimosoid genera by the presence of a combination of characters. In addition to having a more or less fleshy pod that does not open to release the seeds when it is mature (indehiscent, Figure 6-2), the pollen of all of the species is released as single

201

grains (Figure 5-7) rather than in clusters of two, four, eight, or more as in the case of most members of the mimosoid subfamily.

The forty-four species in the genus are distributed in arid and semiarid areas of North and South America, northern Africa, and eastern Asia. Only three species occur naturally in Asia (one of these ranges into northern Africa) and only one additional species is restricted to Africa. Forty of the forty-four species, therefore, are New World natives. North America is the natural "home" of nine of these species, two of which also occur in South America (and one in the West Indies). Thirty-one species are thus indigenous to South America and thirty-five species naturally occur there.

The great morphological diversity of the South American species which encompasses almost all of the characters found in species in North America and the Old World (Burkart, 1976), and the pattern of flavonoid chemistry (Carman, 1973), suggest a South American origin for the genus. Certainly, the process of speciation and ecological diversification has proceeded to a greater extent in extra-tropical South America than elsewhere. The presence in the genus both of distinctive groups of species (sections) and of very similar species within these sections that frequently hybridize indicate *Prosopis* is an old genus which diverged early into several principal lineages, but that within some of these lineages recent episodes of isolation have produced partial speciation.

In the most comprehensive treatment of the genus as a whole, Burkart (1976) recognized five distinctive groups of species formally recognized as sections.

Section *Prosopis*. Three Old World species with spines in the form of prickles along the stems (A-23), round, black fruits (A-24), and glabrous (hairless) petals and ovaries.

Section *Anonychium* Bentham. One Old World species without spines, long flattened fruits, and glabrous petals but hairy ovaries.

Section *Strombocarpa* Bentham. Nine New World species with paired spines formed by modified stipules (A-17) and either coiled fruits (A-19, A-20) with the seeds within them arranged end to end, or short, round fruits (A-22) in which the seeds are stacked. The inside of the flower petals and the ovaries in this section are both hairy (Figure 5-5).

Section *Monilicarpa* Burkart. One New World species with solitary axillary spines (A-14), a bright red fruit with pronounced beading (A-2), and flowers with hairy petals and ovaries.

Section *Algarobia* de Candolle. Thirty New World species with paired (A-1) or solitary (A-14) axillary spines or spiny branch tips (A-13, A-15), variable fruits, but the pods always slightly flattened and the seeds arranged end to end. The inside of the petals and the ovaries are always hairy (Figure 5-5). The North American species of the section, which are among the most morphologically confusing in the genus, have been treated by Johnston (1962).

ILLUSTRATED KEY TO THE SPECIES OF *PROSOPIS**

1a. (p. 213) Plants spiny (if spineless and New World, see 47b) but the spines never scattered along the stem. Spines paired (A–1, A–17) or solitary and nodal (A–14) or formed by the branch tips and segmented (A–15) or striate (A–13). Flower petals hairy near the tip of the inside and the ovary always hairy (Figure 5–5). Inflorescence in the form of a spike (A–5, A–16) or ball (A–21). Pods slightly flattened, beaded, arched, or coiled.

2a. (p. 212) Spines beige or brown, arising above a leaf cluster (A–1) or solitary and axillary (A–14) or formed by branch tips (A–13, A–15). Inflorescence a spike or a slender catkin (A–5, A–16). Flowers white (cream-colored and turning dull yellow when dry) or red. Pods straight (A–10, A–12) or loosely spiraled (Figure 1–11) usually somewhat flattened. Seeds in the pod arranged end to end and separated by a tough, stony endocarp (Figure 6–2).

3a. (below) Pod beaded (A–2) bright red. Spines short, one per axil (A–14). Inflorescence with fewer than 50 flowers (A–16). COMMON NAME. Algarobilla. DISTRIBUTION. Western Argentina in the provinces of Catamarca, La Rioja, and Mendoza.
P. argentina Burkart

3b. Pod straight, arched, or loosely coiled, slightly flattened or round in cross-section, brown, blackish-brown or mottled (A–10, A–12, Figure 1–11). Inflorescences with more than 50 flowers (A–5). Trees and shrubs, sometimes without leaves.

4a. (p. 211) Leafy trees and shrubs. Spines axillary, either paired (A–1) or solitary (A–14) never appearing segmented. Pods flattened to various degrees, straight or twisted.

5a. (below) Leaflets very large and broad between 2 and 10 cm. long and 7 to 32 mm. wide, ovate in outline (A–3). Spines solitary, up to 33 cm. long (A–14). COMMON NAME. Venal, chamacoco. DISTRIBUTION. Eastern Bolivia, southern Paraguay, north central Argentina, and a small area of northeastern Brazil.
P. ruscifolia Grisebach

5b. Leaflets narrower than 5 mm., usually less than 2 cm. long, if longer than 2 cm., narrow in outline (A–6, A–7).

6a. (p. 211) Leaflets 10 to 63 mm. long (rarely as short as 7 mm.), leaflets pairs opposite on the rachis. Inflorescence a long, slender catkin. Spines paired or solitary.

7a. (p. 204) Pods straight, without visible beading (A–12).

A–1
(X 1/3)

A–2
(X 1/5)

A–3
X 1/5

A–4
(X 1/5)

*This key has been modified from Burkart, 1976. An initial following a number means part of a couplet and the number in parenthesis gives the page of the second part of that couplet. Below, means the two parts of a couplet follow one another.

8a. (p. 204) Leaflets oblong, less than 5 times as long as wide, soft (A–6). Pods straight with parallel margins, yellow or brown.

9a. (below) Trees 8 to 20 m. tall. Leaves 2 to 3 times shorter than the inflorescence. Pods straight and 'yellow, 5 to 9 mm. thick and 10 (rarely 6) to 25 cm. long. COMMON NAME. Algarrobo, huarango. DISTRIBUTION. Native to the western dry parts of Colombia, Ecuador, and Perú. Introduced in Puerto Rico, the Hawaiian Islands, the Marquesas, and in parts of Brazil, India, and Australia.

P. pallida (Humboldt & Bonpland ex Willd.) H. B. K.

9b. Trees 3 to 12 m., often shrubby. Leaves usually as long as the inflorescences. Pod 4 to 8 mm. thick and 8 to 29 mm. long, yellow or brown. COMMON NAME. Many, but most commonly, mesquite or algarrobo. DISTRIBUTION. Dry areas of Mexico, particularly southern Mexico, Central America, many of the islands of the West Indies, northern Venezuela and Colombia. Introduced into Brazil.

P. juliflora (Swartz) de Candolle var. *juliflora*

8b. Leaflets linear, over 5 times as long as broad, tough (A–7). Pod straight, brown. COMMON NAME. Honey mesquite, mesquite, (Figure 3–2). DISTRIBUTION. Southwestern U.S.A. in western Texas, New Mexico, eastern Arizona, Oklahoma, Nevada, and Idaho. Northeastern Mexico. Cultivated in Puerto Rico, parts of Asia and Australia. *P. glandulosa* J. Torrey var. *glandulosa*

7b. Pod slightly beaded, usually speckled (Figure 1–11).

10a. (p. 210) Leaves well developed, trees obviously leafy during the growing season (large and full), if small, with more than 3 or 4 pairs of pinnae. Twigs and spines smooth, not striate.

11a. (below) Leaflets shorter than the inflorescences with 3 to 8 (usually 5) pairs of pinnae. Leaflets very close together. Small trees 5 to 6 m. tall. Styles and stigmas bright red. COMMON NAME. Espinillo. DISTRIBUTION. Northeastern Paraguay and neighboring areas of Brazil in the Mato Grosso.

P. rubriflora E. Hassler

11b. Leaves with 1 to 4 pairs of pinnae. Flowers pale white or cream colored with similarly colored styles and stigmas. Dwarf, prostrate, highly branched thorny shrubs, or trees less than 5 m. tall.

A–5 (× 1/5)

A–6 (× 1/3)

A–7 (× 1/3)

12a. (below) Dwarf multi-branched shrubs 10 to 50 cm. tall. Leaves with 1 to 2 pairs of pinnae. Leaflets pubescent, 1.5 to 6 mm. long, linear, 6 to 17 per pinna. Flowers hairy. Inflorescences stubby. COMMON NAME. Algarrobillo. DISTRI-BUTION. Central Argentina, primarily in the mountains of the provinces of Córdoba and San Luis.
P. campestris Grisebach

12b. Trees or large shrubs 1 to 5 m. tall. Pinnae 1 to 4 per leaf. Flowers glabrous on the outside.

13a. (p. 209) Leaves expanded, lax, usually longer than the inflorescence.

14a. (p. 207) Leaflets 4 to 63 mm. long, linear or oblong in outline, not ovate.

15a. (p. 205) Leaflets pubescent, or if glabrous over 10 mm. long and separated from one another (A–4), 5 to 15 mm. long and 2 to 4.8 mm. wide, 1 to 4 pair per pinna.

16a. (below) Leaflets with conspicuous spaces between them (A–4), 12 to 14 pair per pinna. Immature fruit glabrous. Twigs black or gray. COMMON NAME. Algarrobo. DIS-TRIBUTION. Northern Perú.
P. juliflora (Swartz) de Candolle var. horrida (Knuth) Burkart.

16b. Leaflets almost, or actually, touching, 12 to 30 pair per pinna. Immature fruit pubescent. Twigs greenish brown. COMMON NAME. Velvet mesquite. DISTRIBUTION. Southern Arizona and adjacent California, fringing into northern Mexico (Figures 1–11, 3–2, A–8).
P. velutina Wooton

15b. Leaflets glabrous or only slightly ciliate along the margins, large, 10 to 63 mm. long, widely separated along the rachis, the distance between them equal to, or greater than, the width of a leaflet.

A–8
(× 1/3)

P. VELUTINA

Continued on next page

17a. (below) Leaflets linear, 11 to 54 mm. long by 1.1 to 5 (rarely 5) mm. wide. Midvein of leaflets visible but green. Pod usually broad, straight or arched, spines always paired (A–9, Figure 1–11). COMMON NAME. Algarrobo de Chile, algarrobo, panta. DISTRIBUTION. Central Chile and Northwestern Argentina in the provinces of Salta, Tucuman, Catamarca, La Rioja, San Juan, Mendoza, San Luis and Córdoba, dry areas scattered in Bolivia and southern Perú.　　　　*P. chilensis* (Molina) Stuntz

17b. Leaflets oblong in outline with the midveins of the leaflets visible and contrasting in color. Fruit always straight. Spines paired, or solitary on some short shoots.

18a. (below) Leaves 10 to 30 mm. long by 2 to 12 mm. wide. Pods slender, curved or straight, flattened, only slightly beaded. Trees. (This species is probably a hybrid between *P. ruscifolia* and *P. alba*, see Chapter 3). COMMON NAME. Vinalillo. DISTRIBUTION. Argentina in the Gran Chaco region and in the Chaco of Paraguay.　　　　*P. vinalillo* Stuckert

18b. Leaves 7 to 63 mm. long, 5 to 15 times as long as broad. Pod usually thickened, yellow or tinged with violet.

19a. (below) Leaflets stiff, coriaceous. Plants often multistemmed and shrubby. COMMON NAME. Honey mesquite (Figure 3–2). DISTRIBUTION. Eastern Arizona, New Mexico, western Texas, California, northern Mexico. Introduced into Saudi Arabia, Pakistan, India, Burma, southern and southwestern Africa.

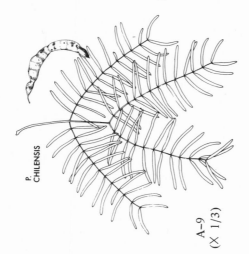

P. CHILENSIS

A–9
(X 1/3)

P. glandulosa J. Torrey (particularly var. *torreyana* (L. Benson) M. C. Johnston

19b. Leaves herbaceous, relatively soft. Usually trees. COMMON NAME. Mesquite, algarrobo. DISTRIBUTION. Coastal areas of Mexico, Baja California, the West Indies and northern South America.

P. juliflora (Swartz) de Candolle

14b. Leaflets elliptic oblong in outline, glabrous, and less than 10(15) mm. long, 12 to 30 per pinna. Pod straight and beaded or with parallel edges, slightly flattened, often curved, yellow or speckled.

20a. (p. 208) Leaflets 15 to 50 per pinna, spaced closely together with distances between them less than the width of a leaf.

21a. (below) Leaflets not conspicuously veined on the underside, 25 to 50 pair per pinna. Pods flattened, often curved or wavy. COMMON NAME. Algarrobo blanco. DISTRIBUTION. Flat areas of subtropical Argentina and adjacent Uruguay and Paraguay and scattered in semiarid areas of Bolivia and Perú.   *P. alba* A. Grisebach

21b. Leaflets conspicuously pinninerved. Fruits straight, thick and yellow, or thick and slightly beaded and pure yellow, speckled, or dark colored.

22a. (below) Pods long, thick, pure yellow in color. Leaflets 15 to 21 pairs per pinna, with very conspicuous pinnate veins on the undersurface. COMMON NAME. Mesquite, algarrobo. DISTRIBUTION. Primarily across the central plateau and hillsides of northern Mexico and southern Texas

Continued on next page

but also disjunct in central and southern Perú, Bolivia, and northern Argentina.

P. laevigata (Humboldt & Bonpland ex Willd.) M. C. Johnston

22b. Pods more or less cylindrical in cross-section, margins with a beaded appearance (A–10) often speckled or darkly colored although sometimes pure yellow. Leaflets 20 to 30 per pinna, obtuse, often slightly ciliate along the margin. The veins underneath in a pinnate pattern, but relatively obscure. COMMON NAME. Algarrobo negro. DISTRIBUTION. Southern Bolivia, most of the northern part of Argentina, Paraguay and western Uruguay.

P. nigra (Grisebach) Hieronymus

20b. Leaflets 60 to 25 pair per pinna, widely spaced with the spaces between them greater than the width of a leaf. Pods always beaded (A–10).

23a. (below) Mat or hummock forming shrubs less than 3 m. tall with underground branching stems. COMMON NAME. Alpataco. DISTRIBUTION. West central Argentina, especially in the creosotebush dominated parts of the central and southern Monte Desert.

P. alpataco R. A. Philippi

23b. Trees without underground branching stems, over 2 m. tall when mature and usually over 3 m.

24a. (below) Spines very conspicuous, 2 to 9 cm. long. COMMON NAME. Algarrobo de las salinas, alpataco. DISTRIBUTION. Central and western Argentina in the provinces of Cordoba, La Rioja, Catamarca, San Juan, and San Luis.   *P. pugionata* Burkart

A–10
(× 1/5)

24b. Spines small (very rarely 3 cm. long) often inconspicuous or even absent (A–11). COMMON NAME. Algarrobo, algarrobo dulce, amarillo, or negro. DISTRIBUTION. Northern Chile (Atacama, Coquimbo) and in the arid regions of western Argentina in the Monte Desert.   *P. flexuosa* de Candolle

13b. Leaves small, shorter than or equal to the length of the inflorescence.

25a. (below) Pod dry, compressed, constricted between the segments (A–11). Leaves with 1 or 2 pair of pinnae and 6 to 20 pairs of leaflets. COMMON NAME. Mesquite amargo. DISTRIBUTION. Northwestern Mexico (Sonora, Baja California) and possibly Arizona.
        P. articulata S. Watson

25b. Pod fleshy, slightly flattened but thick with no constrictions between the segments (edges of the fruit parallel), curved or straight (A–12).

26a. (p. 210) Leaflets with conspicuous nerves on the underside (A–7), 6 to 30 pair per pinna, pubescent or ciliate along the margins. Pods straight with parallel or almost parallel margins.

27a. (p. 210) Pods broad, 9 to 16 mm. in width.

28a. (below) Leaflets obtuse or very rapidly tapering to a narrow point, 15 to 60 pairs per pinna, 2.5 to 8.3 mm. long and 1.4 to 4 mm. broad. Spines small (or absent). Pods straight with almost no constrictions between the segments, pure yellow. COMMON NAME. Algarrobo. DISTRIBUTION. Native in Perú, Colombia, and Ecuador. Introduced into Puerto Rico, the Hawaiian Islands, India, and Australia.
        P. pallida (Humboldt and Bonpland ex Willd.) H.B.K.

Continued on next page

A–11
(× 1/3)

P. FLEXUOSA

A–12
(× 1/8)

28b. Leaflets gradually narrowing to a thick point, linear, 2 to 7.5 mm. long, 0.8 to 1.8 mm. broad. Spines small. Pod with some slight constrictions between the segments, mottled in color. COMMON NAME. Ñandubay, espinillo. DISTRIBUTION. Paraguay, eastern Argentina north of the province of Buenos Aires, Brazil southwest of Rio Grande do Sul, and western Uruguay. Possibly in extreme southern Bolivia.   *P. affinis* Sprengel

27b. Pods narrow, 6 to 8 mm. wide.

29a. (below) Spines strong, up to 8.5 (11.5) cm. long, thicker than supporting branches, often constricted at the base. Pod straight, tinged with violet. Spiny shrubs. COMMON NAME. Algarrobito. DISTRIBUTION. Paraguay and Argentina in the region of the Gran Chaco.   *P. elata* (Burk.) Burkart

29b. Spines weak. Leaves glabrous. Pods straight, straw-yellow colored. Small trees. COMMON NAME. Mesquite. DISTRIBUTION. Mexico in the states of Tamaulipas and Veracruz.   *P. tamaulipana* Burkart

26b. Leaflets without conspicuous nerves on the underside or only the midvein prominent (A–6). Leaves small, delicate. Pods thin, flattened, curved. COMMON NAME. Caldén. DISTRIBUTION. South central Argentina.
                              P. caldenia Burkart

10b. Leaves very small giving the plants an almost leafless appearance. Twigs and spines longitudinally striate (A–13).

30a. (below) Subshrubs with very striate branches and spines. Flowers bright red. Leaves with one pair of pinnae. Leaflets 1 to 3 per pinna. COMMON NAME. Algarobilla, barba de tigre. DISTRIBUTION. Endemic to the pampas and semi-open woodlands of central Argentina.
                              P. humilis Hooker and Arnott

A–13
(natural size)

30b. Trees and large shrubs. Branches and spines only faintly striate. Flowers dull cream-colored. Leaves always with 2 pairs of pinnae. Leaflets 8 to 17 per pinna. COMMON NAME. none. DISTRIBUTION. Paraguay in the northern Chaco. *P. rojasiana* Burkart

6b. Leaflets small, 1.5 to 12 mm. long, alternate on the lower part of the rachis. Inflorescence spikelike or subglobose. Spines solitary (A-14).

31a. (p. 211) Leaflets widely spaced on the rachis. Inflorescence a spike.

32a. (p. 211) Twigs smooth, becoming red soon after formation. Pods fat, round in cross-section, becoming black at maturity. Leaves glabrous.

33a. (below) Leaves less than 8 mm. long with 1 to 4 pairs of leaflets per pinna. Pod partially curved or forming a complete circle or more in one plane. COMMON NAME. Algarrobo patagonico. DISTRIBUTION. Endemic to southern Argentina in the area of Patagonia. *P. denudans* Bentham

33b. Leaves 8 to 9 cm. long with 3 to 9 pairs of leaflets per pinna. Pods thick, straight to slightly curved. COMMON NAME. Unknown. DISTRIBUTION. Only in the province of Mendoza in western Argentina. *P. ruizleali* Burkart

32b. Twigs with longitudinal nerves and remaining green after formation. Pods flattened, olive brown. Leaflets hairy, 4 to 7 per pinna. COMMON NAME. Unknown. DISTRIBUTION. Restricted to the provinces of Mendoza and Neuquen in western Argentina. *P. castellanosii* Burkart

31b. Leaflets close together on the rachis, almost overlapping. Flowers in subglobose heads. COMMON NAME. Cusqui. DISTRIBUTION. Argentina in the mountainous areas of the province of San Juan. *P. calingastana* Burkart

4b. Leafless trees or shrubs (when mature, the juveniles sometimes with leaves). Branch tip forming spines and these "spines" consequently with numerous leaf scars giving the spines a segmented appearance (A-15).

34a. (below) Hummock forming shrubs. Inflorescences resembling a stubby bottle brush (A-16). Twigs slender, 0.8 to 1.3 mm. in diameter. Pods dark and reddish black, 0.8 to 1.3 cm. thick. COMMON NAME. Temoj, barba de tigre. DISTRIBUTION. Widely distributed in the Gran Chaco and Monte regions of Argentina. *P. sericantha* Hooker and Arnott

Continued on next page

A-14 (× 1/2)

A-15 (× 1/10)

A-16 (× 1/2)

34b. Trees with slender, elongate racemes. Twigs over 2.5 mm. thick. Pods almost black, slightly flattened, 1.5 to 2.6 cm. wide. COMMON NAME. Itin, palo mataco. DISTRIBUTION. The Gran Chaco area of northern Paraguay, southern Bolivia and northern and central Argentina.
P. kuntzei Harms

2b. Spines white or yellow, arising from underneath a leaf cluster and formed by stipules that have become spiny (A–17). Inflorescences in round heads (A–21) or in open, stubby spikes (A–16). Pods coiled in several planes (A–19, A–20) or short, stubby, arched (A–18). Leaves always with only one pair of pinna and small leaflets. New World.

35a. (p. 213) Spines white, thin. Seeds in the pod arranged end to end. Pods pointed and arched (A–18) or coiled (A–19, A–20).

36a. (p. 212) Inflorescence an open spike of yellow flowers (A–16).

37a. (below) Pod smooth, slightly flattened, when mature, the seeds scarcely separated from one another (A–18). COMMON NAME. Palo de hierro, palo fierro. DISTRIBUTION. Endemic to Baja California, California.
P. palmeri S. Watson

37b. Pods coiled at least 1 full circle and usually more (A–19, A–20).

38a. (below) Pod lumpy, coiled completely at least once and often more but the coils being very open (A–19). COMMON NAME. Tintitaco, Quenti. DISTRIBUTION. The northwestern provinces of Argentina.
P. torquata (Lagasca) de Candolle

38b. Pods smooth, tightly coiled (see A–20) the area between the seeds delicate.

39a. (below) Pods with 1 to 3 coils. Several pods held in a single cluster. COMMON NAME. none. DISTRIBUTION. Chile in the province of Tarapacá.
P. burkartii Muñoz (probably a hybrid between *P. tamarugo* and *P. strombulifera*)

39b. Pods with numerous coils (A–20) borne singly or only a few in a cluster. COMMON NAME. Screwbean, tornillo. DISTRIBUTION. Southwestern North America, Baja California, and northern Mexico.
P. pubescens Bentham

36b. Inflorescences globose (A–21) or ovoid in shape, less than 1 cm. in diameter.

40a. (below) Pods red-brown with 2 to 5 irregular coils. Inflorescence ovoid and with rather densely clustered flowers. Trees or shrubs. COMMON NAME. Algarrobillo espinoso. DISTRIBUTION. West-central Argentina. Rather rare.
P. abbreviata Bentham

A–17
(X 2/3)

A–18
(X 1/3)

A–19
(X 1/3)

A–20
(X 2/5)

40b. Pods yellow with more than 5 irregular coils. Inflorescence a round ball (A–21). Small creeping shrubs forming thickets.

41a. (below) Leaflets widely spaced along the rachis, 3 to 8 pairs per pinna, glabrous. COMMON NAME. Retortuño, espinilla. DISTRIBUTION. Western Argentina and northern Chile.
P. strombulifera (Lamarck) Bentham

41b. Leaflets close together, almost overlapping, usually 6 to 12 pair per pinna, glabrous or variously pubescent.

42a. (below) Leaflets slightly pubescent. Bracts small. COMMON NAME. Mastuerzo, retortuño. DISTRIBUTION. Central Perú in the province of Huancavelica and in central Argentina.
P. reptans Bentham var. reptans

42b. Leaflets glabrous or pubescent. Bracts large and hairy. COMMON NAME. Screwbean. DISTRIBUTION. U.S.A. in Texas and adjacent areas of Mexico.
P. reptans var. cinerascens (A. Gray) Burkart

35b. Spines yellow, large. Seeds in the pod arranged horizontally, side to side. Pods straight or curved (A–22) short (less than 5 cm).

43a. (below) Spines 0.5 to 3.8 cm. long. Leaflets blunt tipped. Fruit curved into one-half or a complete circle, dark brown. Seeds 3 to 4.3 mm. long. COMMON NAME. Tamarugo. DISTRIBUTION. Northern Chile in the province of Tarapacá.
P. tamarugo F. Philippi

43b. Spines up to 5 cm. long. Leaflets with acute tips. Pods straight or slightly bent, brown orange, short (A–22). Seeds 4.5 to 6 mm. long. COMMON NAME. Churqui. DISTRIBUTION. Southern Bolivia and the most northern parts of Argentina in the puna.
P. ferox Grisebach

1b. Inside of the petals hairless. Branches without spines or with short prickles along the branches (A–23). Stipules either falling off early or leafy. Natives of the Old World.

44a. (p. 214) If spiny, the spines appearing as scattered prickles (A–23). Stipules leafy, but falling off. Ovary glabrous. Anther gland straight.

45a. (below) Leaves with 2 to 7 pairs of leaflets. Pods borne on a short stalk, round in cross-section. Shiny, smooth black seeds arranged in two rows with horizontal separations. COMMON NAME. Acatin. DISTRIBUTION. Throughout the Middle East in Algeria, Tunesia, Egypt, Turkey, Cyprus, Syria, Israel, Iran, Afganistan, Pakistan, Transcaucasia, and the USSR.
P. farcta (Russell) Macbride

Continued on next page

A–21
(X 3/4)

A–22
(X 1/2)

A–23
(natural size)

45b. Leaves with 1 to 3 pair of leaflets. Pods borne on a stalk 0.8 to 2 cm. long, slender, straight, containing one row of seeds arranged end to end.

46a. (below) Pods very slender and long, 8 to 9 cm. long and 4 to 7 mm. wide. Seeds in the fruit conspicuously separated from one another. Inflorescences often borne in a cluster subtended by a bract. COMMON NAME. Jand, ghaf, shumi. DISTRIBUTION. The Middle East in Saudi Arabia, Iran, Afganistan, Pakistan, and India.

P. cineraria (Linnaeus) Druce

46b. Pods short, 4 to 6 cm. long and 8 to 10 mm. wide. Seeds slightly overlapping in the pod. Inflorescences borne singly and with no bract at the base. COMMON NAME. none. DISTRIBUTION. Restricted to the desert of Iran.

P. koelziana Burkart

44b. Trees without spines. Ovary hairy. Stipules small.

47a. (below) Pod round in cross-section, black (A–24), the seeds free inside so that the pod produces a rattle when shaken. Gland on the anthers of the flowers broad and on an inward curving stalk. Old World. COMMON NAME. variable. DISTRIBUTION. Central Africa in Senegal, Guinea Bissau to Nigeria, the Cameroons, Sudan, Uganda, and Ethiopia.

P. africana (Guillemin, Perrottet and Richard) Taubert

A–24
(X 1/5)

47b. Pod slightly flattened, often with a beaded appearance (A–8). The seeds firm inside the endocarp (Figure 6–2) usually yellow or speckled. Anther glands small (Figure 5–5). New World. (These taxa are in some cases varieties of species which have spines.)

48a. (p. 214) Leaflets large, 0.8 to 6.5 cm. long. Branch tips reddish. Leaves darkening when dry, containing 1 pair of pinnae.

49a. (below) Leaflets 2 to 9 per pinna, ovate in outline, 2.5 to 6.5 cm. long by 2 to 4 cm. wide. Pod straight flattened, yellow. COMMON NAME. none. DISTRIBUTION. The Chaco of Paraguay and Formosa in Argentina.

P. fiebrigii Harms

49b. Leaflets 6 to 26 per pinna, linear in outline, 0.8 to 4 cm. long and 1.8 to 16 mm. wide. Pod flattened, yellow. COMMON NAME. algarrobo. DISTRIBUTION. Paraguay and Argentina in the eastern Chaco.

P. hassleri Harms

48b. Leaflets less than 1.3 cm. long, linear. Twigs and branch tips green. Leaves with 1 or 2 pair of pinnae.
50a. (p. 215) Leaflets glabrous.

51a. (below) Leaflets widely spaced on the rachis, 5 to 13 cm. long. COMMON NAME. Algarrobo dulce, or negro. DISTRIBUTION. Western Argentina in the provinces of Tucuman and San Juan.

P. flexuosa de Candolle var. *subinermis* Burkart

51b. Leaflets practically touching, 1.5 to 3.2 mm. long. COMMON NAME. Algarrobo negro. DISTRIBUTION. Argentina, this form primarily in the province of Santa Fe.   *P. nigra* (Grisebach) Hieronymus var. *ragonesei* Burkart

50b. Leaflets pubescent.

52a. (below) Leaflets 2.5 to 8.3 mm. long. Inflorescences longer than the leaves. COMMON NAME. Algarrobo. DISTRIBUTION. Native to Perú, Colombia and Ecuador. Introduced into Puerto Rico, Hawaii, Australia, the Marquesas Islands, and India.

P. pallida (Willd.) H. B. K.

52b. Leaflets 6 to 20 mm. long. Inflorescences equal in length to, or shorter than, the leaves. COMMON NAME. Algarrobo, mesquite. DISTRIBUTION. This variety is found in Ecuador, the species extends throughout northern South America, dry areas of Central America to Mexico, the West Indies.

P. juliflora (Swartz) de Candolle var. *inermis* (H. B. K.) Burkart

References

Alcoze, T. M., and E. G. Zimmerman. 1973. Food habits and dietary overlap of two heteromyid rodents from the mesquite plains of Texas. J. Mammal. *54*:900-908.

Anderson, J. O., and C. E. Dibble. 1950-1959. Florentine Codex, general history of the things of New Spain (Translation of the manuscript by B. Sahagun, 1575). 7 Vols. Sante Fe, New Mexico.

Anonymous. 1857. Texas almanac. Dallas Morning News, Dallas, Texas.

Aparicio, F. de. 1946. The Comechingón and their neighbors of the Sierras de Córdoba, pp. 673-686. *In* J. H. Steward (ed.), Handbook of South American Indians. Smithsonian Institution. Bur. Am. Ethnol. Bull. 143. Vol. 2. The Andean civilizations. 1035 pp.

Ares, F. N. 1974. The Jornada Experimental Range. Soc. Range Manage. Monogr. I. Denver, Col. 74 pp.

Arnold, L. W. 1942. Notes on the life history of the sand pocket mouse. J. Mammal. *23*:339-341.

Aschmann, H. 1959. The central desert of Baja California. Demography and ecology. Ibero-Americana 42. Univ. Calif. Press., Berkeley, Calif. 282 pp.

Atchison, E. 1951. Studies in the Leguminosae. VI. Chromosome numbers among tropical woody species. Am. J. Bot. *38*:538-546.

Baegert, J. J. 1772. Nachrichten von der Amerikanischen Halbinsel Californien: mit einem zwe jachen anhang jalscher Nachrichten. Manhiem, Germany. (Translated by M. M. Brandenburg and C. L. Baumann. Berkeley, Calif. 1952). 218 pp.

Bailey, H. 1977. Current climate, pp. 25-49. *In* G. H. Orians and O. T. Solbrig (eds.), Convergent evolution in warm deserts. Dowden, Hutchinson & Ross, Stroudsburg, Pa. 352 pp.

Balls, E. K. 1962. Early uses of California plants. Univ. Calif. Press, Berkeley, Calif. 103 pp.

Bandelier, A., and F. Bandelier. 1905. The journey of Alvar Nuñez Cabeza de Vaca and his companions from Florida to the Pacific 1528-1536. A. S. Barns, New York. 231 pp.

Baquar, S. R., A. Husain, and S. Akhtar. 1966. Meiotic chromosome numbers in some vascular plants of Indus Delta II. Bot. Not. *119*:26-32.

Barrows, D. P. 1967. The ethnobotany of the Cahuilla Indians of Southern California. Including a Cahuilla bibliography and introductory essays by H. W. Lawton, L. J. Bean, and W. Bright. Malki Mus. Press, Banning, California. 82 pp.

Bartlett, J. R. 1854. Personal narrative of explorations and incidents in Texas,

New Mexico, California, Sonora and Chihuahua, etc. 2 Vols. Appleton, New York. 505 & 624 pp.

Bateman, G. C. 1967. Home range studies of a desert nocturnal rodent fauna. Ph.D. thesis. Univ. Ariz., Tucson, Arizona. 115 pp.

Bean, L. J., and K. S. Saubel. 1961. Cahuilla ethnobotanical notes: the aboriginal use of oak, pp. 237-345. *In* Archaeological Survey Annual Report. Univ. Calif., Los Angeles, Calif. 446 pp.

Bean, L. J., and K. S. Saubel. 1963. Cahuilla ethnobotanical notes: the aboriginal uses of mesquite and screwbean, pp. 55-74. *In* Archeological Survey Annual Report. Univ. Calif., Los Angeles, Calif. 446 pp.

Bean, L. J., and K. S. Saubel. 1972. Temalpakh. Malki Mus. Press. Banning, Calif. 225 pp.

Bell, E. A. 1962. Associations of ninhydrin-reacting compounds in the seeds of 49 species of *Lathyrus.* Biochem. J. *83*:225-229.

Bell, E. A., and A. S. L. Tiramanna. 1965. Associations of amino acids and related compounds in the seeds of forty-seven species of *Vicia*: their taxonomic and nutritional significance. Biochem. J. *97*:104-111.

Bell, W. H., and E. F. Castetter. 1937. Ethnobiological studies in the American Southwest V. The utilization of mesquite and screwbean by the aborigines in the American Southwest. Univ. New Mexico Biol. Ser. *5*(2):1-55.

Berlin, B. 1973. Folk systematics in relation to biological classification and nomenclature. Annu. Rev. Ecol. Syst. *4*:259-271.

Bir, S. S., and S. Sidhu. 1967. Cytological observations on the north Indian members of family Leguminosae. The Nucleus *10*:47-63.

Bird, J. 1948. Preceramic cultures in Chicama and Virú. Soc. Am. Archaeol. Memoir *4*:21-28.

Blair, W. F. 1941. Annotated list of mammals of the Tularosa Basin, New Mexico. Am. Midl. Nat. *26*:218-229.

Blair, W. F. 1943a. Ecological distribution of mammals in the Tularosa Basin, New Mexico. Univ. Mich. Contr. Lab. Vert. Biol. *20*:1-24.

Blair, W. F. 1943b. Populations of the deer-mouse and associated small mammals in the mesquite associations of southern New Mexico. Univ. Mich. Contr. Lab. Vert. Biol. *21*:1-40.

Bleckmann, C. A., and H. M. Hull. 1975. Leaf and cotyledon surface ultrastructure of five *Prosopis* species. J. Ariz. Acad. Sci. *10*:98-105.

Bogusch, E. R. 1950. A bibliography on mesquite. Texas J. Sci. 1950:528-537.

Bohrer, V. L. 1962. Nature and interpretation of ethnobotanical materials from Tonto National Monument, pp. 75-114. *In* L. R. Caywood (ed.), Archaeological studies at Tonto National Monument, Arizona. Southwestern Monuments Ass. Tech. Ser. 2. 176 pp.

Bohrer, V. L. 1970. Ethnobotanical aspects of Snaketown, a Hoholsam village in southern Arizona. Am. Antiquity *35*:413-430.

Bohrer, V. L. 1973a. Ethnobotany of Point of Pines Ruin, Arizona. W:10:50. Econ. Bot. *27*:423-437.

Bohrer, V. L. 1973b. Tularosa Valley Project. Tentative list of utilized plant remains from Fresnal Shelter, January, 1972, pp. 211–218. *In* Human Systems Research, Tech. Manual: 1973 survey of the Tularosa Basin. Human Systems Res., Inc., Albuquerque, New Mexico. 495 pp.

Bolton, H. E. 1919. Father Escobar's relation of the Oñate Expedition to California. Catholic Hist. Rev. *5*:19–41.

Bolton, H. E. 1930. Anza's California expeditions. Vol. 3 Univ. Calif. Press, Berkeley, Calif. 436 pp.

Bourke, J. G. 1889. Notes on the cosmogony and theogony of the Mojave Indians of the Rio Colorado, Arizona. J. Am. Folk-Lore *2*:169–189.

Bourke, J. G. 1894. Popular medicines, customs, and superstitions of the Rio Grande. J. Am. Folk-Lore *7*:119–146.

Bowen, T. 1976. Seri prehistory, the araeology of the central coast of Sonora. Anthropol. Pap. Univ. Ariz. 27, 120 pp.

Bowen, T., and E. Moser. 1968. Seri pottery. The Kiva *33*:89–132.

Bradley, W. G., and R. A. Mauer. 1973. Rodents of a creosotebush community in southern Nevada. Southwest. Nat. *17*:333–344.

Bray, W. L. 1904. Forest resources of Texas. U.S.D.A. Bull. 47. 71 pp.

Bray, W. L. 1906. Distribution and adaptation of the vegetation of Texas. Univ. Texas Sci. Ser. *10*:73–74.

Brewer, F. G. 1970. An introduction to isozyme techniques. Academic Press, New York. 150 pp.

Bridwell, J. C. 1918. Notes on the Bruchidae and their parasites in the Hawaiian Islands. Proc. Hawaiian Entomol. Soc. *3*:465–509.

Bridwell, J. C. 1920. Insects injurious to the algaroba feed industry. Hawaiian Planters Rec. *22*:337–343.

Brock, J. H. 1972. Edaphic factors as affected by honey mesquite canopies. Proc. 25th Meeting Soc. Range Manage. *25*:8.

Bryant, V. M. 1974. Prehistoric diet in southwest Texas: the caprolite evidence. Am. Antiquity *39*:407–420.

Buffington, L. C., and C. H. Herbel. 1965. Vegetational changes on a semidesert grassland range from 1958 to 1963. Ecol. Monogr. *35*:139–164.

Bull, L. B., C. C. J. Culvenor, and A. T. Dick. 1968. The pyrrolizidine alkaloids, pp. 1–293. *In* A. Neuberger and E. L. Tatum (eds.), Frontiers of biology. American Elsevier, New York. 293 pp.

Burkart, A. 1937. Estudios morfológicos y biológicos en el género *Prosopis*. Darwiniana *3*:27–48.

Burkart, A. 1940. Materiales para una monografía del género *Prosopis* (Leguminosae). Darwiniana *4*:57–128.

Burkart, A. 1943. Las leguminosas argentinas. Acme, Buenos Aires, Argentina. 590 pp.

Burkart, A. 1952. Las leguminosas argentinas silvestres y cultivadas. Acme, Buenos Aires, Argentina. 569 pp.

Burkart, A. 1976. A monograph of the genus *Prosopis* (Leguminosae subfam. Mimosoideae). J. Arnold Arb. *57*(3):219–249; 1976, J. Arnold Arb. *57*(4):450–455.

Burt, W. H., and R. P. Grossenheider. 1964. A field guide to the mammals. Houghton Mifflin, Boston, Mass. 284 pp.

Cabrera, A., and J. Yepes. 1940. Mamíferos sudamericanos (vida, costumbres y descripción). Compannia Argentina de Editores, Buenos Aires, Argentina. 370 pp.

Cabrera, P. P. 1929. Los aborígenes del país de Cuyo. Rev. Univ. Nac. Córdoba 15 and 16. pp. 1–396.

Callen, E. O. 1969. Diet as revealed by caprolites, pp. 186–194. In D. R. Brothwell and E. Higgs (eds.), Science in archaeology. A comprehensive survey of progress and research. Ed. I. Basic Books, New York. 595 pp.

Camp, B. J., and M. J. Norvell. 1966. The phenylethylamine alkaloids of native range plants. Econ. Bot. 20:274–278.

Carlson, G. G., and V. H. Jones. 1940. Some notes on uses of plants by the Comanche Indians. Pap. Mich. Acad. Sci. Arts Letters 25:517–542.

Carman, N. J. 1973. Systematic and ecological investigations in the genus Prosopis (Mimosaceae) emphasizing the natural products chemistry. Ph.D. thesis, Univ. Texas, Austin, Texas. 221 pp.

Casanova, E. 1946. The cultures of the Puna and the Quebrada de Humahuaca, pp. 619–632. In J. H. Steward (ed.), Handbook of South American Indians. Smithsonian Institution. Bur. Am. Ethnol. Bull. 143. Vol. 2. The Andean Civilizations. 1035 pp.

Castetter, E. F., and R. M. Underhill. 1935. The ethnobiology of the Papago Indians. Univ. New Mexico Bull. 275. Ethnobotanical Studies of the American Southwest 4(3). 84 pp.

Castetter, E. F., and M. E. Opler. 1936. III. The ethnobiology of the Chiricahua and Mescalero Apache. A. The uses of plants for foods, beverages and narcotics. Univ. New Mexico Bull. 297. Ethnobotanical Studies of the American Southwest 4(5). 63 pp.

Castetter, E. F., and W. H. Bell. 1937. The aboriginal utilization of the tall cacti in the American southwest. Univ. New Mexico Bull. 307. Ethnobotanical Studies in the American Southwest 5(1). 48 pp.

Castetter, E. F., and W. H. Bell. 1951. Yuman Indian agriculture. Primitive subsistence on the lower Colorado and Gila Rivers. Univ. New Mexico Press, Albuquerque, New Mexico. 274 pp.

Cates, R. G., and G. H. Orians. 1975. Successional status and the palatability of plants to generalized herbivores. Ecology 56:410–418.

Cates, R. G., G. H. Orians, D. F. Rhoades, J. Schultz, and C. S. Tomoff. 1977. The plant-foliage-eater-predator-system, pp. 166–196. In G. H. Orians and O. T. Solbrig (eds.), Convergent evolution in warm deserts. Dowden, Hutchinson & Ross, Stroudsburg, Pa. 352 pp.

Cates, R. G., and D. F. Rhoades. 1977. Patterns in the production of antiherbivore chemical defenses in plant communities. Biochem. Syst. in press.

Center, T. D., and C. D. Johnson. 1974. Coevolution of some seed beetles (Coleoptera: Bruchidae) and their hosts. Ecology 55:1096–1103.

Cherubini, C. 1954. Número de cromosomas de algunas especies del género *Prosopis* (Leguminosae-Mimosoideae). Darwiniana *10*:637–643.

Chew, R. M. 1961. Ecology of the spiders of a desert community. J. New York Entomol. Soc. *69*:5–41.

Chew, R. M., and B. B. Butterworth. 1964. Ecology of rodents in Indian Cove (Mojave Desert), Joshua Tree National Monument, California. J. Mammal. *45*:203–225.

Chew, R. M., and A. E. Chew. 1970. Energy relationships of the mammals of a desert shrub (*Larrea tridentata*) community. Ecol. Monogr. *40*:1–21.

Chittenden, N. H. 1901. Among the Cocapahs. Land of Sunshine *14*:196–204.

Clavigero, F. J. 1789. Storia della California. Appresso Modesto Fenzo, Venice. (Translated as The History of (Lower) California by S. E. Labe, and A. A. Gray, Stanford Univ. Press, Stanford, California. 1937.) 413 pp.

Clotts, H. V. 1915. Report on nomadic Papago surveys, submitted to C. R. Olberg, Supt. of Irrigation, U.S. Indian Service. Unpub. manuscript. Arizona State Mus. Archives. 107 pp.

Cobo, P. B. 1653. Historia del Nuevo Mundo. *3*:56. Sevilla 1890–1895.

Cockerell, T. D. A. 1900. Some bees visiting the flowers of mesquite. Entomologist *33*:243–245.

Cocucci, A. E. 1965. Estudios en el género *Prosopanche* (Hydnoraceae). Kurtziana *2*:53–74.

Cook, O. F. 1908. Change of vegetation on the south Texas prairies. U.S.D.A. Bur. Plant Industry Circ. *14*:1–7.

Cosgrove, C. B. 1947. Caves of the upper Gila and Hueco areas in New Mexico and Texas. Pap. Peabody Mus., Harvard Univ. *24*:1–181.

Covas, G. 1950. Número de cromosomas en seis dicotiledóneas argentinas. Bol. Soc. Arg. Bot. *3*:83.

Covas, G., and B. Schnack. 1946. Número de cromosomas en Antófitas de la región de Cuyo (República Argentina). Rev. Arg. Agron. *13*:153–166.

Covas, G., and B. Schnack. 1947. Estudios cariológicos en Antófitas II. Rev. Arg. Agron. *14*:224–231.

Culin, R. S. 1907. Games of the North American Indians. Bur. Am. Ethnol. Rep. 24 (for 1903). 846 pp.

Curtin, L. S. M. 1949. By the prophet of the earth. San Vicente Foundation, Sante Fe, New Mexico. 160 pp.

Curtin, L. S. M. 1965. Healing herbs of the upper Rio Grande. Southwest Museum, Los Angeles, Calif. 281 pp.

Curtis, E. S. 1908. The North American Indians. Vol. 2. Harvard Univ. Press, Cambridge, Mass. 142 pp.

Cushman, R. A. 1911. Notes of the host plants and parasites of some North American Bruchidae. J. Econ. Entomol. *4*:489–510.

Dahl, B. E., R. E. Sosebee, and J. P. Green. 1973. Research high lights. Texas Tech. Univ. *4*:1–19.

Darlington, C. D. 1939. The evolution of genetic systems. Ed. I. Cambridge Univ. Press, Cambridge, England. 149 pp.

Densmore, F. 1932. Yuman and Yaqui music. Bur. Am. Ethnol. Government Printing Off., Washington, D.C. 216 pp.

Dixon, A. F. G. 1973. Biology of aphids. Edward Arnold, London, England. 58 pp.

Dunnill, P. M., and L. Fowden. 1967. The amino acids of the genus *Astragalus*. Phytochem. *6*:1659-1663.

Earle, F. R., and Q. Jones. 1962. Analysis of seed samples from 113 plant families. Econ. Bot. *16*:221-250.

Easter, S. J., and R. E. Sosebee. 1975. Influence of soil-water potential on the water relationships of honey mesquite. J. Range Manage. *28*:230-232.

Eilberg, B. A. 1973. Presencia de diseminulos de "vinal" (*Prosopis ruscifolia* Griseb.) en deyecciones de equinos y bovinos. Ecología (Buenos Aires) *1*:56-57.

Eisenberg, J. F. 1963. The behavior of heteromyid rodents. Univ. Calif. Publ. Zool. *69*:1-114.

Emmart, E. W. (ed.). 1940. The Badianus Manuscript, an Aztec herbal of 1552 by M. de la Cruz and J. Badianus. Johns Hopkins Univ. Press, Baltimore, Md. 341 pp.

Euler, R. C., and V. H. Jones. 1956. Hermetic sealing as a technique of food preservation among the Indians of the American Southwest. Proc. Am. Phil. Soc. *100*:87-99.

Fahn, A. 1974. Plant anatomy. Ed. 2. Pergamon Press, New York. 611 pp.

Feeny, P. 1970. Seasonal changes in oakleaf tannins and nutrients as a cause of spring feeding by wintermoth caterpillars. Ecology *51*:656-681.

Feldman, I. 1972. Consideraciones sobre algunas problemas de lenosas invasoras en la Republica Argentina malezas y su control. Organo Official Ass. Arg. Para el Control de Malezas. Ano *1*(3):25-27.

Felger, R. S. 1975. Nutritionally significant new crops for arid lands: a model from the Sonoran Desert, pp. 373-403. *In* J. Mayer and J. W. Dyer (eds.), Priorities in child nutrition. UNESCO. Vol. 2. 404 pp.

Felger, R. S. 1976. Investigaciones ecologicas en Sonora y localidades adyacentes en Sinaloa: una perspectiva, pp. 21-62. *In* B. Braniff and R. S. Felger (eds.), Sonora: anthropoligía del desierto. Inst. Nac. Anthropol. Hist. Mexico. Mexico City, Mexico. 592 pp.

Felger, R. S., and M. B. Moser. 1971. Seri use of mesquite (*Prosopis glandulosa* var. *torreyana*). The Kiva *37*:53-60.

Felger, R. S., and M. B. Moser. 1973. Eelgrass (*Zostera marina* L.) in the Gulf of California: discovery of its nutritional value by the Seri Indians. Science *181*:355-356.

Felger, R. S., and M. B. Moser. 1974a. Columnar cacti in Seri culture. The Kiva *39*:257-356.

Felger, R. S., and M. B. Moser. 1974b. Seri Indian Pharmacopoeia. Econ. Bot. *28*:414-436.

Felger, R. S., and M. B. Moser. 1976. Seri Indian food plants: desert subsistence without agriculture. Ecol. Food and Nutrition *5*:13-27.

Fewkes, J. W. 1912. Casa Grande, Arizona. Bur. Ethnol. Rep. 28 (for 1907):25–179.

Fisher, C. E. 1947. Present information on the mesquite problem. Texas Agric. Exp. Sta. PR-1056. 7 pp.

Fisher, C. E. 1950. The mesquite problem in the Southwest. J. Range Manage. *3*:60–70.

Fisher, C. E. 1962. Evaluation of brush and pasture improvement practices in Argentina. Spec. Rep. Prepared for U.S. AID, State Dept., Washington, D.C. 34 pp.

Fisher, C. E. 1964. Control of brush and grassland improvement in the Chaco Region of Paraguay. Res. Rep. for U.S. AID, State Dept., Washington, D.C. in coop. with Fac. Agron. Vet., Univ. Paraguay and Texas Agric. Exp. Sta. 30 pp.

Fisher, C. E., J. L. Fultz, and H. Hope. 1946. Factors affecting action of oils and water soluble chemicals in mesquite eradication. Ecol. Monogr. *16*:109–126.

Fisher, C. E., C. H. Meadors, R. Behrens, E. D. Robison, P. T. Marion, and H. L. Morton. 1959. Control of mesquite on grazing lands. Texas Agric. Exp. Sta. Bull. *935*:1–23.

Fisher, C. E., H. T. Wiedemann, J. P. Walter, C. H. Meadors, J. H. Brock, and B. T. Cross. 1972. Brush control research on rangeland. Texas Agric. Exp. Sta. MP-1043. 18 pp.

Fisher, C. E., G. O. Hoffman, and C. J. Scifres. 1973a. The mesquite problem. Univ. Texas Agric. Exp. Sta. Res. Monogr. *1*:5–9.

Fisher, C. E., H. T. Wiedemann, C. H. Meadors, and J. H. Brock. 1973b. Mechanical control of mesquite. Texas Agric. Exp. Sta. Res. Monogr. *1*:46–52.

Flannery, K. V. 1968. Archaeological systems theory and early Mesoamerica, pp. 67–68. *In* B. J. Meggers (ed.), Anthropological archaeology in the Americas. Anthropol. Soc. Wash., Washington, D.C. 151 pp.

Fleming, T. H. 1973. Numbers of mammal species in North and Central American forest communities. Ecology *54*:555–563.

Fontana, B. L., W. J. Robison, C. W. Cormack, and E. E. Leavitt, Jr. 1962. Papago Indian pottery. Univ. Wash. Press, Seattle, Wash. 163 pp.

Forbes, R. H. 1895. The mesquite tree: its products and uses. Ariz. Exp. Sta. Bull. *13*:1–12.

Forde, C. D. 1931. Ethnography of the Yuma Indians. Univ. Calif. Publ. Am. Archaeol. Ethnol. *28*:83–278.

Forister, G. 1970. Bionomics and ecology of 11 species of Bruchidae (Coleoptera). M.S. thesis, Northern Ariz. Univ., Flagstaff, Ariz. 93 pp.

Fosberg, F. R. 1966. Miscellaneous notes on Hawaiian plants –4. Bishop Mus. Occas. Pap. *23*:129–138.

Gates, D. M. 1968. Transpiration and leaf temperatures. Annu. Rev. Plant Physiol. *19*:211–238.

Gentry, H. S. 1942. Rio Mayo plants. Carnegie Inst. Washington Publ. *527*:1–328.

Gentry, H. S. 1963. The Warihio Indians of Sonora–Chihuahua: an ethnographic survey. Bur. Am. Ethnog. Bull. *186*:61-144.

Gifford, E. W. 1918. Clans and moieties in southern California. Univ. Calif. Publ. Am. Archaeol. Ethnol. *14*:155-219.

Gifford, E. W. 1931. The Kamia of Imperial Valley. Bur. Am. Ethnol. Bull. *97*:1-94.

Gifford, E. W. 1932. The Southeastern Yauapa. Univ. Calif. Publ. Am. Archaeol. Ethnol. *29*:176-251.

Gifford, E. W. 1933. The Cocopa. Univ. Calif. Publ. Am. Archaeol. Ethnol. *31*:257-334.

Gifford, E. W. 1936. Northeastern and western Yavapi. Univ. Calif. Publ. Archaeol. Ethnol. *34*:247-354.

Glendening, C. E., and H. A. Paulsen, Jr. 1955. Reproduction and establishment of velvet mesquite as related to invasion of semidesert grasslands. U.S.D.A. Tech. Bull. 1127. 50 pp.

Gottlieb, L. D. 1973. Genetic differentiation, sympatric speciation and the origin of a diploid species of *Stephamomeria*. Am. J. Bot. *60*:545-553.

Graham, J. D. 1960. Morphological variation in mesquite (*Prosopis*, Leguminosae) in the lowlands of northeastern Mexico. Southwest. Nat. *5*:187-193.

Grant, V. 1958. The regulation of recombination in plants. Cold Springs Symposia Quant. Biol. *23*:337-363.

Grant, V. 1971. Plant speciation. Columbia Univ. Press, New York. 435 pp.

Graziano, M. N., G. E. Ferraro, and J. D. Coussio. 1971. Alkaloids of Argentine medicinal plants. II. Isolation of turmine, β-phenethylamine and tryptamine from *Prosopis alba*. Lloydia *34*:453-454.

Greegor, D. 1974. Comparative ecology and distribution of two species of armadillo, *Chaetophractus vellerosus* and *Dasypus novemcinctus*. Ph.D. thesis. Univ. Arizona, Tucson, Ariz. 114 pp.

Griffen, W. B. 1959. Notes on Seri Indian culture, Sonora, Mexico. Monogr. School of Inter-American Stud. 10. Univ. Florida Press, Gainesville, Fla. 54 pp.

Griffiths, D. 1904. Range investigations in Arizona. U.S.D.A. Bur. Plant Industry Bull. *67*:1-62.

Grossman, F. E. 1873. The Pima Indians of Arizona. Smithsonian Institution Rep. (for 1871). 14 pp.

Haas, R., and J. Dodd. 1972. Water-stress patterns in honey mesquite. Ecology *53*:674-680.

Haas, R. H., R. E. Meyer, C. J. Scifres, and J. H. Brock. 1973. Growth and development of mesquite. Texas Agric. Exp. Sta. Res. Monogr. *1*:10-19.

Halevy, G. 1974. Effects of gazelles and seed beetles (Bruchidae) on germination and establishment of *Acacia* species. Israel J. Bot. *23*:120-126.

Hammond, G. P., and A. Rey (eds.). 1940. Narratives of the Coronado Expedition 1540-1542. Publ. Coronado Cuarto Centennial 1540-1940. Vol. II. 413 pp. Univ. New Mexico Press, Albuquerque, New Mexico.

Harborne, J. B. (ed.). 1972. Phytochemical ecology. Academic Press, New York. 272 pp.

Harrington, M. R. 1933. Gypsum Cove, Nevada. Southwest Mus. Pap. *8*:1-197.

Hastings, J. R., and R. M. Turner. 1965. The changing mile. Univ. Ariz. Press, Tucson, Ariz. 317 pp.

Haury, E. 1945. The excavation of Los Muertos and neighboring ruins in the Salt River Valley, southern Arizona. Pap. Peabody Mus. Harvard Univ. *24*:1-223.

Haury, E. 1950. The stratigraphy and archaeology of Ventana Cave, Arizona. Univ. Ariz. Press, Tucson, Ariz. 599 pp.

Haury, E. 1975. The Hohokam: desert farmers and craftsmen, excavations at Shaketown, 1964-1965. Univ. Ariz. Press, Tucson, Ariz. 412 pp.

Havard, V. 1895. Food plants of the North American Indians. Torrey Bot. Club Bull. *22*:98-123.

Havard, V. 1896. Drink plants of the North American Indians. Torrey Bot. Club Bull. *23*:33-46.

Hayden, J. D. 1967. A summary prehistory and history of the Sierra Pinacate, Sonora. Am. Antiquity *32*:335-344.

Hayden, J. D. 1969. Gyratory crushers of the Sierra Pinacate, Sonora. Am. Antiquity *34*:154-161.

Herbel, C. H. 1964. Brush control the cattleman. *51*:160.

Heredia, O. R. 1968. Arqueología de la subárea de las selvas occidentales. Actas Memorias, 37th Internac. Congr. Americanistas *2*:295-253.

Hicks, F. 1961. Ecological aspects of aboriginal culture in the western Yuman area. Ph.D. thesis. Univ. Calif., Los Angeles, Calif.

Hinckley, A. D. 1960. The klu beetle, *Mimosestes sallaei* (Sharp) in Hawaii (Coleoptera: Bruchidae). Proc. Hawaiian Entomol. Soc. *17*:260-269.

Holden, W. C., C. C. Seltzer, R. F. Studhalter, C. J. Wagner, and W. G. McMillan. 1936. Studies of the Yaqui Indians of Sonora, Mexico. Texas Tech. College Bull. *12*:3-142.

Hole, F., F. V. Flannery, and J. A. Neeley. 1969. Prehistory and human ecology of the Deh Luran Plain: an early village sequence from Khuzistan, Iran. Mus. Anthropol. Univ. Mich., Ann Arbor, Memoir I. 504 pp.

Hooper, L. 1920. The Cahuilla Indians. Univ. Calif. Publ. Am. Archaeol. Ethnol. *16*:316-380.

Hrdlička, A. 1906. Notes on the Pima of Arizona. Am. Anthropol. *8*:39-46.

Hrdlička, A. 1908. Physiological and medical observations among the Indians of southwestern United States and northern Mexico. Bur. Am. Ethnol. Bull. *34*:1-460.

Hubby, J. L., and R. C. Lewontin. 1966. A molecular approach to the study of genic heterozygosity in natural populations. I. The number of alleles at different loci in *Drosophila pseudoobscura*. Genetics *54*:577-594.

Hull, H. M., S. J. Shellhorn, and R. E. Saunier. 1971. Variations in creosotebush (*Larrea divaricata*) epidermis. J. Ariz. Acad. Sci. *6*:196-205.

Hull, H. M., H. Morton, and J. Wharrie. 1975. Environmental influences on cuticle development and resultant foliar penetration. Bot. Rev. *41*:421–452.

Humboldt, A. von. 1806. Ideen zu einer Physiognomick der Gewächse. Tubingen, Germany. 26 pp.

Humphrey, R. R. 1949. Fire as a control of undesirable shrubs. J. Range Manage. *2*:175–182.

Hunziker, J. H., L. Poggio, C. A. Naranjo, R. A. Palacios, and A. B. Andrade. 1975. Cytogenetics of some species and natural hybrids in *Prosopis* (Leguminosae). Canadian J. Genetics Cytol. *17*:253–262.

Isley, D. 1972. Legumes of the U.S. VI. *Calliandra, Pithecellobium, Prosopis.* Madroño *21*:273–297.

Ives, R. L. 1964. The Pinacate Region, Sonora, Mexico. Occas. Pap. Calif. Acad. Sci. *47*:1–43.

Jain, S. K. 1969. Comparative ecogenetics of two *Avena* species occurring in central California. Evol. Biol. *3*:73–118.

Janzen, D. H. 1969. Seed-dispersal versus seed size, number, toxicity and dispersal. Evolution *23*:1–27.

Janzen, D. H. 1971a. The fate of *Scheelea rostrata* fruits beneath the parent tree: predispersal attack by bruchids. Principes *15*:89–101.

Janzen, D. H. 1971b. Escape of juvenile *Dioclea megacarpa* (Leguminosae) vines from predators in a deciduous tropical forest. Am. Nat. *105*:97–112.

Janzen, D. H. 1971c. Seed predation by animals. Annu. Rev. Ecol. Syst. *2*:465–492.

Janzen, D. H. 1972. Escape of *Cassia grandis* L. beans from predators in time and space. Ecology *52*:964–979.

Johnston, M. C. 1962. The North American mesquites. *Prosopis* section *Algarobia* (Leguminosae). Brittonia *14*:72–90.

Johnston, M. C. 1963. Past and present grasslands of southern Texas and northeastern Mexico. Ecology *44*:456–466.

Jones, D. A. 1972. Cyanogenic glycosides and their function, pp. 103–122. *In* J. B. Harborne (ed.), Phytochemical ecology. Academic Press, New York. 272 pp.

Jones, V. 1941. The plant materials from Winona and Ridge Ruin. Appendix No. 2, pp. 295–300. *In* J. C. McGregor. Winona and Ridge Ruin. Part I. Mus. N. Ariz. Univ. Bull. *18*. Flagstaff, Ariz. 309 pp.

Jones, Q., and F. R. Earle. 1966. Chemical analysis of seeds II: oil and protein content of 759 species. Econ. Bot. *20*:127–155.

Kannan, K. 1923. The function of the prothoracic plate in mylabrid (bruchid) larvae. Mysore State Dept. Agric. Entomol. Ser. Bull. *7*:1–47.

Kingsolver, J. M. 1964. The genus *Neltumius* (Coleoptera: Bruchidae). Coleop. Bull. *18*:105–111.

Kingsolver, J. M. 1967. On the genus *Ripibruchus* Bridwell, with descriptions of a new species and a closely related new genus. Proc. Entomol. Soc. Wash. (D.C.) *69*:318–327.

Kingsolver, J. M. 1968. A new genus of Bruchidae from South America, with the description of a new species. Proc. Entomol. Soc. Wash. (D.C.) *70*:280-286.

Kingsolver, J. M. 1972. Description of a new species of *Algarobius* Bridwell (Coleoptera: Bruchidae). Coleop. Bull. *26*:116-120.

Kissell, M. L. 1916. Basketry of the Papago and Pima. Am. Mus. Nat. Hist. Anthropol. Pap. *17*:115-264.

Kroeber, A. L. 1931. The Seri. Southwest Mus. Pap. *6*:1-60. Southwest Mus., Los Angeles, Calif.

Lacher, J. R., J. Davies, K. Gardner, and C. Jones. 1963. Germination studies on mesquite seeds. Proc. N. Am. Ass. Office Seed Analysis *53*:58-66.

Lamprey, H. F., G. Halevy, and S. Makacha. 1974. Interactions between *Acacia*, bruchid seed beetles and large herbivores. East African Wildl. J. *12*:81-85.

Latcham, R. E. 1936. La agricultura precolumbiana en Chile y Ios países vecinos. Univ. Chile, Santiago, Chile. 336 pp.

LeClaire, J., and G. Brown. 1974a. Summary of qualitative phenology data—Andalgalá, Catamarca, Argentina. July 2, 1973-January 21, 1974. Origin and Structure of Ecosystems Tech. Rep. 74-9. 37 pp.

LeClaire, J., and G. Brown. 1974b. Summary of qualitative phenology data—Andalgalá, Catamarca, Argentina. January 28-March 25, 1974. Origin and Structure of Ecosystems Tech. Rep. 74-16. 18 pp.

LeClaire, J., P. Reppun, and P. Cantino. 1973a. Summary of qualitative phenology data—Andalgalá, Catamarca, Argentina. September 1, 1972-January 31, 1973. Origin and Structure of Ecosystems Tech. Rep. 73-2. 38 pp.

LeClaire, J. P. Reppun, and P. Cantino. 1973b. Summary of qualitative phenology data—Andalgalá sites, Argentina. January 22, 1973-May 21, 1973. Origin and Structure of Ecosystems Tech. Rep. 73-19. 37 pp.

León-Portilla, M. 1973. Historia natural y crónica de la antigua California (de Miguel del Barco). Univ. Nac. Autonems de Mexico, Inst. Invest. Hist. de Mexico. Mexico City, Mexico. 464 pp.

Levins, R. 1968. Evolution in changing environments. Princeton Univ. Press. Princeton, New Jersey. 120 pp.

Levitt, J. 1972. Response of plants to environmental stresses. Academic Press, New York. 697 pp.

Linsley, E. G. 1958. The ecology of solitary bees. Hilgardia *27*:541-599.

Long, A., R. M. Hansen, and P. S. Martin. 1974. Extinction of the Shasta ground sloth. Geol. Soc. Am. Bull. *85*:1843-1848.

Long, S. H. 1820. An account of an expedition from Pittsburgh to the Rocky Mountains performed in the years 1819 and 1820. From the notes of Major Long, Mr. Say and others of the exploring party, Edwin James Corp, pp. 1-51. *In* R. A. Thwaites (ed.), Early western travels 1784-1846. Vols. 14-17.

Lothrop, S. 1946. Indians of the Paraná Delta and La Plata littoral, pp. 177-190. *In* J. H. Steward (ed.), Handbook of South American Indians. Smithsonian Institution. Bur. Am. Ethnol. Bull. 143. Vol. I. The marginal tribes. 624 pp.

Lovell, J. H. 1926. Honey plants of North America (north of Mexico). A. I. Root, Medina, Ohio. 408 pp.

Lowe, C. H., J. H. Morello, J. K. Cross, and G. Goldstein. 1977. Gradient Analysis of Bajada Communities, pp. 92–100. *In* G. H. Orians and O. T. Solbrig (eds.), Convergent evolution in warm deserts. Dowden, Hutchinson & Ross, Stroudsburg, Pa. 352 pp.

Lumbreras, L. G. 1974. The peoples and cultures of Perú. Translated by B. J. Meggers. Smithsonian Institution, Washington, D.C. 248 pp.

Lumholtz, C. 1912. New trails in Mexico. Scribner, New York. 411 pp.

Mabry, T. J., J. Hunziker, and D. R. DiFeo, Jr. 1977. Creosote Bush: Biology and chemistry of *Larrea* in New World deserts. Dowden, Hutchinson & Ross, Stroudsburg, Pa. 368 pp.

MacArthur, R. H., and J. W. MacArthur. 1961. On bird species diversity. Ecology *42*:594–598.

McGee, W. J. 1898. The Seri Indians. Bur. Am. Ethnol. Ann. Rep. *17*:1–344.

McGee, W. J. 1971. The Seri Indians. Reprinted from Bur. Am. Ethnol. Ann. Rep. *17*:1–344 (1898) Rio Grande Press, Glorieta, New Mexico. 406 pp.

McMillan, C., and J. T. Peacock. 1964. Bud-bursting in diverse populations of mesquite (*Prosopis*: Leguminosae) under uniform conditions. Southwest. Nat. *9*:181–188.

MacNeish, R. S. 1964. The food-gathering and incipient agriculture stage of prehistoric Middle America, pp. 413–426. *In* R. C. West (ed.), Natural environments and early cultures. Handbook of Middle American Indians. Vol. I. Univ. Texas Press, Austin, Texas. 570 pp.

Malin, J. C. 1953. Soil, animal and plant relationships of the grassland historically reconsidered. Sci. Monthly *76*:207–220.

Manje, J. M. 1926. Luz de tierra incognita en la América septentrional y diario de la exploraciónes en Sonora. Publ. Archiv. General de la Nación. Vol. 10., F. del Castillo (ed.). English translation of part 2 by H. T. Kerns and associates, entitled: Unknown Arizona and Sonora 1693–1721. Ariz. Silhouettes, Tucson, Ariz. 303 pp.

Marcy, R. B. 1849. Report of the exploration and surveys of the route from Fort Smith, Arkansas to Sante Fe, New Mexico made in 1849. House EX Document 45, 31st Congress of the U.S. Pub. Doc. 577, Washington, D.C.

Mares, M. A. 1973. Climates, mammalian communities and desert rodent adaptations: an investigation into evolutionary convergence. Ph.D. thesis. Univ. Texas, Austin, Texas. 345 pp.

Mares, M. A. 1975. Observations of Argentine desert rodent ecology, with emphasis on water relations of *Eligmodontia typus*, pp. 155–175. *In* I. Prakash and P. K. Ghosh (eds.), Rodents in desert environments. W. Junk, The Hague, Netherlands. 624 pp.

Mares, M. A., and A. C. Hulse. 1977. Patterns of some vertebrate communities in creosotebush deserts. *In* T. S. Mabry, J. Hunziker and D. R. DiFeo, Jr. (eds.), Creosote Bush: biology and chemistry of *Larrea* in New World deserts. Dowden, Hutchinson & Ross, Stroudsburg, Pa. 368 pp.

Markert, C. L. 1968. The molecular basis for isozymes. Ann. N.Y. Acad. Sci. *151*:14–40.

Márquez Miranda, F. 1946. The Chaco-Santiagueño culture, pp. 655–660. *In* J. H. Steward (ed.), Handbook of South American Indians. Smithsonian Institution. Bur. Am. Ethnol. Bull. 143. Vol. 2. The Andean Civilizations. 1035 pp.

Martin, P. S. 1967. Prehistoric overkill, pp. 75–120. *In* P. S. Martin and H. E. Wright, Jr. (eds.), Pleistocene extinctions: the search for a cause. Yale Univ. Press, New Haven, Conn. 453 pp.

Martin, S. C. 1948. Mesquite seeds remain viable after 44 years. Ecology *29*:393.

Martin, S. C., and F. S. Tshirley. 1962. Mesquite seeds live a long time. Prog. Agric. in Ariz. *14*:15.

Martin, S. C., and D. R. Cable. 1974. Managing grass and shrub ranges. U.S.D.A. Forest Service Tech. Bull. 1480. 45 pp.

Martinez, M. 1959. Plantas útiles de la flora mexicana. Ed. 3. Ediciones Botas, Mexico City, Mexico. 621 pp.

Martinez Crovetto, R. 1965. Estudios ethnobotánicos I. Nombres de plantas y su utilidad según los indios tobas del este del Chaco. Bonplandia *1*:279–333.

Martinez Crovetto, R. 1968. Estado actual de las tribus Mocovíes del Chaco (R.A.). Ethnobiológica *12*:1–25.

Mather, K. 1943. Polygenic inheritance and natural selection. Biol. Rev. *18*:32–64.

Mathiot, M. 1973. A dictionary of Papago usage. Vol. I. 504 pp. Indiana Univ. Press, Bloomington, Ind.

Meigs, P. 1939. The Kiliwa Indians of Lower California. Ibero Americana *15*:1–114.

Meinzer, O. E. 1927. Plants as indicators of ground water. U.S. Geol. Survey Water Supply Pap. 577. 95 pp.

Merck. 1960. Merck index of chemicals and drugs. Ed. 7. Merck, Rahway, New Jersey. 1642 pp.

Merrill, R. E. 1923. Plants used in basketry by the California Indians. Univ. Calif. Publ. Archaeol. Ethnol. *20*:213–242.

Métraux, A. 1929. Contribution a l'ethnographie et l'archaeologie de la Province de Mendoza (R.A.). Rev. Inst. Ethnol. Univ. Nac. Tucuman *1*(1):1–75.

Métraux, A. 1946. Ethnography of the Chaco, pp. 197–370. *In* J. A. Steward (ed.), Handbook of South American Indians. Smithsonian Institution. Bur. Am. Ethnol. Bull. 143. Vol. I. The marginal tribes. 624 pp.

Meyer, R. E., J. L. Morton, R. H. Haas, E. D. Robison, and T. E. Riley. 1971. Morphology and anatomy of honey mesquite. Texas Agric. Exp. Sta. Tech. Bull. 1423. 186 pp.

Moldenke, A. R., and J. L. Neff. 1974. The bees of California: a catalogue with special reference to pollination and ecological research. Origin and Structure of Ecosystems Tech. Reps. 74-1:245 pp., 74-2:41 pp., 74-3:288 pp., 74-4:257 pp., 74-5:189 pp., 74-6:53 pp.

Monson, G. 1943. Food habits of the banner-tailed kangaroo rat at Arizona. J. Wildl. Manage. 7:98-102.

Monson, G., and W. Kessler. 1940. Life history notes on the banner-tailed kangaroo rat, Merriam's kangaroo rat, and the white-throated wood rat in Arizona and New Mexico. J. Wildl. Manage. 4:37-43.

Mooney, H. A. 1972. The carbon balance of plants. Annu. Rev. Ecol. Syst. 3:315-346.

Mooney, H. A., and E. L. Dunn. 1970. Convergent evolution of Mediterranean-climate evergreen sclerophyll shrubs. Evolution 24:292-303.

Mooney, H. A., O. Björkman, and J. Berry. 1975. Photosynthetic adaptations to high temperatures, pp. 138-151. In N. Hadley (ed.), Environmental physiology of desert organisms. Dowden, Hutchinson & Ross, Strouds-burg, Pa. 283 pp.

Morello, J. H. 1972. The *Prosopis ruscifolia* woodland in the Chaco region. Origin and Structure of Ecosystems Newsletter 2(3):A3.

Morello, J. H., N. E. Crudelli, and M. Saraceno. 1971. Los vinalares de Formosa. In La vegetación de la Republica Argentina. INTA Publ. Ser. Fito-geogr. 11. 113 pp.

Moser, E. 1963. Seri bands. The Kiva 28:14-27.

Moser, E. 1973. Seri basketry. The Kiva 38:105-140.

Moser, E. 1976. Guia bibliográfica de las fuentes para el estudio de la ethno-grafía Seri, pp. 365-375. In B. Braniff and R. S. Felger (eds.), Sonora, antropología del desierto. Collect. Cient. Diversa 27. Inst. Nac. Antropol. Hist. Centro Regional del Noreste. Mexico City, Mexico. 592 pp.

Moser, M. B. 1970. Seri from conception through infancy. The Kiva 35:201-210.

Muñoz P., C. 1971. Una nueva especie de *Prosopis* para el norte de Chile. Bol. Mus. Nac. Hist. Nat., Chile 32:363-370.

Neff, J. L., B. B. Simpson, A. R. Moldenke. 1977. Flowers-flower visitor system, pp. 204-224. In G. H. Orians and O. T. Solbrig (eds.), Convergent evolution in warm deserts. Dowden, Hutchinson & Ross, Stroudsburg, Pa. 352 pp.

Nentvig, J. 1971. Descripción geografica de Sonora. (Ed. by German Viveros). Publ. Arch. Gen. de la Nacion. Ser 2 (1). 247 pp. Mexico City, Mexico.

Norris, J. J., K. A. Valentine, and J. B. Gerard. 1963. Mesquite control with monuron, fenuron and diuron. New Mexico Agric. Exp. Sta. Bull. 484. 14 pp.

Olsen, R. W. 1973. Shelter-site selection in the white-throated woodrat, *Neotoma albigula*. J. Mammal. 54:594-610.

Orians, G. H., and O. T. Solbrig (eds.). 1977. Convergent evolution in warm deserts. Dowden, Hutchinson & Ross. Stroudsburg, Pa. 352 pp.

Oviedo y Valdes, G. F. de. 1535. Historia natural y general de las indias. Vol. 9. Part 3. Sevilla, Spain.

Owen, R. C. 1963. The use of plants and non-magical techniques in curing illness among the Paipai, Santa Catalina, Baja California, Mexico. Am. Indigena 23:319-344.

Palmer, E. 1878. Plants used by the Indians of the United States. Am. Nat. 12:593-606 and 646-655.

Parker, K. W., and S. C. Martin. 1952. The mesquite problem on southern Arizona ranges. U.S.D.A. Circ. 908. 70 pp.

Pascual R. 1970. Evolución de communidades cambios faunisticos e integraciones biocenoticas de los vertebrados cenozoicos de Argentina. Actas IV Congr. Latinoam. Zool., Caracas, Venezuela 1968. 2:991-1088.

Patterson, B., and R. Pascual. 1968. The fossil mammal fauna of South America. Quat. Rev. Biol. 43:409-451.

Peacock, J. T., and C. McMillan. 1965. Ecotypic differentiation in Prosopis (mesquite). Ecology 46:35-51.

Percival, M. S. 1961. Types of nectar in angiosperms. New Phytol. 60:235-281.

Pfaffenberger, G. S. 1974. The biosystematics of first instar larvae of some North American Bruchidae. Ph.D. thesis, Northern Ariz. Univ., Flagstaff, Ariz. 203 pp.

Pfaffenberger, G. S., and C. D. Johnson. 1976. Biosystematics of the first-stage larvae of some North American Bruchidae (Coleoptera). U.S.D.A. Tech. Bull. 1525:1-75.

Phillips, W. S. 1963. Depths of roots in soil. Ecology 44:424.

Pyykkö, M. 1966. The leaf anatomy of east Patagonian xeromorphic plants. Ann. Bot. Fenn. 3:453-622.

Reynolds, H. G., and H. S. Haskell. 1949. Life history notes on rice and barley pocket mice of southern Arizona. J. Mammal. 30:150-156.

Reynolds, H. G., and G. E. Glendening. 1949. Merriam kangaroo rats a factor in mesquite propagation on southern Arizona rangeland. J. Range Manage. 2:193-197.

Reynolds, H. G., and G. E. Glendening. 1950. Relation of kangaroo rats to range vegetation in southern Arizona. Ecology 31:456-463.

Rhoades, D. F. 1977. The anti-herbivore defences of Larrea. In T. J. Mabry, J. Hunziker, and D. R. DiFeo (eds.), Creosote Bush: biology and chemistry of Larrea in New World deserts. Dowden, Hutchinson & Ross, Stroudsburg, Pa. 368 pp.

Rhoades, D. F., and R. G. Cates. 1976. Toward a general theory of plant anti-herbivore chemistry, pp. 168-213. In J. W. Wallace and R. L. Mansell (eds.), Biochemical interactions between plants and insects. Rec. Adv. Phytochem. 10. Plenum Press, New York. 425 pp.

Robelo, C. A. 1948. Diccionario de azteqismos, o sea jardin de las raises aztecas; palabras del idioma nahuatl, azteca, baja diversas formas. Contr. Dictionario Nac. Ed. 3. Ediciones Fuente Cultural, Mexico City, Mexico. 548 pp.

Robison, E. D. 1967. New development in mesquite control. Rolling Plains Livestock Res. Sta. Tech. Rep. 5:1–4.

Robison, E. D. 1968. Rate of reinfestation and survival of mesquite seedlings on native grassland. Texas Agric. Exp. Sta. PR 2584. 3 pp.

Robison, E. D., B. T. Cross, and P. T. Marion. 1970. Beef and forage production following honey mesquite control in the Texas Rolling Plains. Texas Agric. Exp. Sta. PR-2803:15–17.

Rockwood, L. L. 1974. Seasonal changes in the susceptibility of Crescentia alata leaves to the flea beetle Oedionychus sp. Ecology 55:142–148.

Root, A. I. 1966. The ABC and XYZ of bee culture. Revised by E. R. Root. Assisted by H. H. Root and J. A. Root. Ed. 33. A. I. Root Co., Medina, Ohio, 712 pp.

Rosenzweig, M. L. 1973. Habitat selection experiments with a pair of co-existing heteromyid rodent species. Ecology 54:111–117.

Rosenzweig, M. L., and J. Winakur. 1969. Population ecology of desert rodent communities, habitats and environmental complexity. Ecology 50:558–571.

Rothschild, M. 1973. Secondary plant substances and warning coloration in insects, pp. 59–83. In H. van Emden (ed.), Insect/plant relationships. Wiley, New York. 215 pp.

Russell, E. 1908. The Pima Indians. Bur. Am. Ethnol. Rep. 26 (for 1904–1905):3–390.

Ryan, R. M. 1968. Mammals of Deep Canyon. Desert Mus., Palm Springs, Calif. 137 pp.

Schimper, A. F. W. 1903. Plant geography upon a physiological basis. Clarendon Press, Oxford, England. 839 pp.

Schmidt-Nielsen, K. 1964. Desert animals. Clarendon Press, Oxford, England. 277 pp.

Schnack, B., and G. Covas. 1947. Estudios cariológicos en Antófitas. I. Haumania 1:32–41.

Schuster, J. L. (ed.). 1969. Literature on mesquite (Prosopis L.) of North America. Texas Tech. Univ. Spec. Rep. No. 26. 84 pp.

Scifres, C. J., and J. H. Brock. 1970. Growth and development of honey mesquite seedlings in the field and greenhouse as related to time of planting, planting depth, soil temperature and top removal. Texas Agric. Exp. Sta. PR-2817:65–71.

Scifres, C. J., and J. H. Brock. 1970. Moisture-temperature interrelations in germination and early seedling development of honey mesquite. Texas Agric. Exp. Sta. PR-2816:63–64.

Scifres, C. J., and J. H. Brock. 1971. Thermal regulation of water uptake by germinating honey mesquite seeds. J. Range Manage. 24:157–158.

Scifres, C. J., and J. H. Brock. 1972. Emergence of honey mesquite seedlings relative to planting depth and soil temperatures. J. Range Manage. 25:217-219.

Scifres, C. J., J. H. Brock, and R. R. Hahn. 1971. Influence of secondary invasion on honey mesquite invasion in north Texas. J. Range Manage. 24:206-210.

Seneviratne, A. S., and L. Fowden. 1968. The amino acids of the genus Acacia. Phytochem. 7:1039-1045.

Šesták, Z., P. G. Jarvis, and J. Čataký. 1971. Criteria for the selection of suitable methods, pp. 1-48. In Z. Šesták and P. G. Jarvis, and J. Čataký (eds.), Plant photosynthetic production (manual of methods). W. Junk, The Hague, Netherlands. 818 pp.

Simpson, B. B. 1977. Breeding systems of dominant perennial plants of two disjunct warm desert ecosystems. Oecologia 27:203-226.

Simpson, B. B., and F. Vervoorst. 1977. Physiolographic Settings of the Study Area. pp. 16-25. In G. H. Orians and O. T. Solbrig (eds.), Convergent evolution in warm deserts. Dowden, Hutchinson & Ross, Stroudsburg, Pa. 352 pp.

Smith, C. E., Jr. 1967. Plant remains. The prehistory of the Tehuacan Valley, pp. 220-255. In D. S. Byers (ed.), Environment and subsistence. Univ. Texas Press, Austin, Texas. 331 pp.

Smith, J. G. 1899. Grazing problems in the southwest and how to meet them. U.S.D.A. Div. Agron. Bull. 16. 47 pp.

Smith, L. L., and D. N. Ueckert. 1974. Influence of insects on mesquite production. J. Range Manage. 27:61-65.

Solbrig, O. T. 1972. Breeding systems and genetic variation in Leavenworthia. Evolution 26:155-160.

Solbrig, O. T., and K. Bawa. 1975. Isozyme variation in species of Prosopis (Leguminosae). J. Arnold Arb. 56:398-412.

Solbrig, O. T., and P. D. Cantino. 1975. Reproductive adaptations in Prosopis (Leguminosae, Mimosoideae). J. Arnold Arb. 56:185-210.

Soo Hoo, C. F., and G. Fraenkel. 1966. The consumption, digestion, and utilization of food plants by a polyphagous insect, Prodenia eridania (Cramer). J. Insect Physiol. 12:711-730.

Spicer, E. H. 1962. Cycles of conquest. Univ. Ariz. Press, Tucson, Ariz. 609 pp.

Spier, L. 1933. Yuman tribes of the Gila River. Univ. Chicago Press, Chicago, Ill. 433 pp.

Standley, P. 1923. Trees and shrubs of Mexico. Contr. U.S. Nat. Herb. 23(2):171-515.

Stebbins, G. L. 1950. Variation and evolution in plants. Columbia Univ. Press, New York. 643 pp.

Steward, J. H. (ed.). 1950. Tribal and linguistic distributions of South America, Map 18. In J. H. Steward (ed.), Handbook of South American Indians. Smithsonian Institution. Bur. Am. Ethnol. Bull. 143. Vol. 6. Physical anthropology, linguistics and cultural geography of South American Indians. 715 pp.

Strain, B. R. 1970. Field measurements of tissue water potential and carbon dioxide exchange in the desert shrubs *Prosopis juliflora* and *Larrea divaricata*. Photosynthetica *4*:118-122.

Strong, W. D., and C. Evans. 1952. Cultural stratigraphy in the Virú Valley, northern Perú: the Formative and Florescent Epochs. Columbia Studies Archaeol. Ethnol. 4. Columbia Univ. Press, New York. 373 pp.

Stuart, B. R. 1943. Paiute surprise the Mohave. Masterkey *27*:218-219.

Stuckert, T. 1900. El vinalillo. Anal. Mus. Nac., Buenos Aires *7*:73-78.

Sudzuki, F. 1969. Absorción foliar de humedad atmospherica en tamarugo, *Prosopis tamarugo* Phil. Bol. Tecnico *30*:1-23. Univ. Chile, Santiago, Chile.

Swenson, W. H. 1969. Comparisons of insects on mesquite in burned and unburned areas. M.S. thesis. Texas Tech. College, Lubbock, Texas. 61 pp.

Swier, S. R. 1974. Comparative seed predation strategies of mesquite bruchids in Arizona with particular reference to seed height, direction and density. Ph.D. thesis. Northern Ariz. Univ., Flagstaff, Ariz. 94 pp.

Tiedemann, A., R. James, O. Klemmedson, and P. R. Ogden. 1971. Response of four perennial Southwestern grasses to shade. J. Range Manage. *24*:442-447.

Toro, H., and C. D. Michener. 1975. The subfamily Xeromelissinae and its occurrence in Mexico (Hymenoptera: Colletidae). J. Kansas Entomol. Soc. *48*:351-357.

Tortorelli, L. A. 1956. Maderas y bosques argentinas. Acme, Buenos Aires. 910 pp.

Towle, M. 1961. The ethnobotany of pre-Colombian Perú. Viking Fund Publ. Anthropol. *30*:55-56.

Treutlein, T. (ed. and translator). 1949. Sonora, a description of the province by I. Pffeferkorn. Coronado Cuarto Centennial Publ. 12. 329 pp. Univ. New Mexico Press, Albuquerque, New Mexico.

Trippel, E. J. 1889. The Yuma Indians. Overland Monthly Ser. 2. *13*:561-584 and *14*:1-11.

Ueckert, D. N. 1973. Effect of leaf-footed bugs on mesquite reproduction. J. Range Manage. *26*:227-229.

Ueckert, D. N., K. L. Polk, and C. R. Ward. 1971. Mesquite twig girdler: a possible means of mesquite control. J. Range Manage. *24*:116-118.

Underhill, R. M. 1939. Social organization of the Papago Indians. Columbia Univ. Press, New York. 280 pp.

Vestal, P. A., and R. E. Schultes. 1939. The economic botany of the Kiowa Indians. Bot. Mus. Harvard Univ., Cambridge, Mass. 110 pp.

Vogel, V. 1970. American Indian medicine. Univ. Okla. Press. Norman, Okla. 583 pp.

Vorhies, C. T., and W. P. Taylor. 1922. Life history of the kangaroo rat *Dipodomys spectabilis spectabilis* Merriam. U.S.D.A. Bull. 1091. 40 pp.

Vorhies, C. T., and W. P. Taylor. 1933. Life history and ecology of jack rabbits, *Lepus alleni* and *Lepus californicus* in relation to grazing in Arizona. Ariz. Agric. Exp. Sta. Tech. Bull. 49. 117 pp.

Vorhies, C. T., and W. P. Taylor. 1940. Life history and ecology of the white-throated wood rat, *Neotoma albigula albigula* Hartley, in relation to grazing in Arizona. Ariz. Agric. Exp. Sta. Tech. Bull. 86. 74 pp.

Waldbauer, G. P. 1964. The consumption, digestion and utilization of solanaceous and non-solanaceous plants by larvae of the tobacco hornworm, *Protoparce sexta* (John.). Entomol. Exp. Appl. 7:253–269.

Walton, G. P. 1923. A chemical and structural study of mesquite, carob, and honey locust beans. U.S.D.A. Bull. 1194. 19 pp.

Wendt, C. W., R. H. Haas, and J. R. Runkles. 1968. Influence of selected environmental variables on the transpiration rate of mesquite (*Prosopis glandulosa* var. *glandulosa* (Torr.) (Cockr.). Am. Soc. Agron. J. 60:382–384.

Went, F. W. 1975. Water vapor absorption in *Prosopis*, pp. 67–75. *In* F. J. Vernberg (ed.), Physiological adaptation to the environment. Intext Publ., New York. 576 pp.

Werner, F. G., and G. D. Butler. 1958. A survey of the insects on mesquite in southern Arizona. Prelim. Rep. (unpubl.) Dept. Entomol. Univ. Ariz. Tucson, Ariz. 9 pp.

Whittaker, R. H., and P. P. Feeny. 1971. Allelochemics: chemical interactions between species. Science 171:757–770.

Winship, G. P. 1896. The Coronado Expedition. Bur. Am. Ethnol. Rep. 14(1):329–637.

Wodehouse, R. P. 1971. Hayfever plants—their appearance, distribution, time of flowering and their role in hayfever. Ed. 2. Habner, New York. 280 pp.

Wright, H. A. 1973. Noxious brush and weed control. Res. Highlights Texas Tech. Univ. 4. 15 pp.

Wood, D. L. 1973. Selection and colonization of ponderosa pine by bark beetles, pp. 101–117. *In* H. F. van Emden (ed.). Insect/plant relationships. Wiley, New York. 215 pp.

Yacovleff, E., and E. Herrera. 1934, 1935. El mundo vegetal del los antiguos peruanos. Rev. Mus. Nac., Lima, Perú, 3 & 4.

Yang, T. W., and Y. Abe. 1973a. Summary of qualitative phenology data—Saguaro National Monument East and West, Arizona, U.S.A. June to November, 1972. Origin and Structure of Ecosystems Tech. Rep. 73-1. 25 pp.

Yang, T. W., and Y. Abe. 1973b. Summary of qualitative phenology data—Saguaro National Monument East and West, Arizona, U.S.A. November 18, 1972–July 13, 1973. Origin and Structure of Ecosystems Tech. Rep. 73-18. 26 pp.

Zacher, F. 1952. Die Nährpflanzen der Samenkäfer. Zeitschr. Angew. Entomol. 33:460–479.

Index of Scientific Names

Included in this index are all of the scientific names that appear in the text, tables or figures. The authors of each taxon and the family to which it belongs are also given. Page citations for *Prosopis chilensis*, *P. flexuosa*, *P. glandulosa*, *P. velutina* and *P. torquata* are not given because these names are found so ubiquitously throughout the book. In addition to these names, synonyms of the currently recognized species of *Prosopis* are also included with an indication of the correct taxon that includes the type of the synonym. There has been much confusion in the application of various names to species of *Prosopis* and it is hoped that the inclusion of synonyms will help non-botanists who must use names for species of *Prosopis*. The nomenclature followed here is that of Burkart 1976.

Subject Index